TOWARD A LEAN AND LIVELY CALCULUS

Conference/Workshop
To Develop Alternative
Curriculum and
Teaching Methods
For Calculus at the College Level
Tulane University, January 2-6, 1986

Editor
RONALD G. DOUGLAS
State University of New York
Stony Brook, New York 11794

List of Participants and Addresses
Sloan Conference/Workshop On Calculus Instruction
January 2–6, 1986 Tulane University

Professor Lida Barrett
Provost's Office
Northern Illinois University
DeKalb, IL 60115

Professor Robert Davis
603 West Michigan Avenue
Urbana, IL 61801

Professor Ronald G. Douglas
Department of Mathematics
State University of New York
Stony Brook, NY 11794

Professor Susanna Epp
Department of Mathematics
DePaul University
Chicago, IL 60614

Professor Andrew Gleason
Department of Mathematics
Harvard University
Cambridge, MA 02138

Dr. Sam Goldberg
Alfred P. Sloan Foundation
630 Fifth Avenue
New York, NY 10111–0242

Professor Jerome Goldstein
Department of Mathematics
Tulane University
New Orleans, LA 70118

Professor John Kenelly
Department of Mathematics
Clemson University
Clemson, SC 29631

Professor Peter Lax
Courant Institute
New York University
New York, NY 10012

Ms. Katherine Layton
Beverly Hills High School
Beverly Hills, CA 90212

Professor Steve Maurer
Department of Mathematics
Swarthmore College
Swarthmore, PA 19081

Professor Louise Raphael
Department of Mathematics
Howard University
Washington, DC 20059

Professor Peter Renz
Department of Mathematics
Bard College
Annandale on Hudson, NY 12504

Professor Stephen Rodi
Department of Mathematics
Austin Community College
Austin, TX 78768

Professor Alan Schoenfeld
School of Education
Tolman Hall
University of California
Berkeley, CA 94720

Professor Donald B. Small
Department of Mathematics
Harvey Mudd College
Claremont, CA 91711

Professor Lynn Steen
Department of Mathematics
St. Olaf's College
Northfield, MN 55057

Professor Sherman Stein
Department of Mathematics
University of California
Davis, CA 95616

Dr. James Stevenson
President's Office
Georgia Institute of Technology
Atlanta, GA 30332

Professor Steven S. Terry
Department of Mathematics
Ricks College
Rexburg, ID 83440

Dr. John Thorpe
National Science Foundation
1800 G Street, NW
Washington, DC 20550

Professor Thomas Tucker
Department of Mathematics
Colgate University
Hamilton, NY 13346

Professor Robert H. Van Der Vaart
Department of Mathematics
North Carolina State University
Raleigh, NC 27695

Dean Mac E. Van Valkenburg
University of Illinois
106 Engineering Hall
1308 West Green Street
Urbana, IL 61801

Professor Paul Zorn
Department of Mathematics
St. Olaf's College
Northfield, MN 55057

© 1986 by
The Mathematical Association of America
Library of Congress Catalog Card Number 86-063683
ISBN Number 0-88385-056-7

Current Printing (Last digit)
6 5 4 3 2 1

Table of Contents

INTRODUCTION

Steps Toward a Lean and Lively Calculus

Calculus is central to the mathematical sciences, is fundamental to the study of all sciences and engineering, and belongs in the core undergraduate mathematics curriculum for all students. The rapid development of modern physical science which began several centuries ago coincided with the invention of the calculus and this symbiotic intermingling grew to include engineering and eventually the biological and social sciences. Mathematics continues to evolve and the collegiate curriculum has evolved with it. Despite these changes, calculus has been a central part of the college curriculum for more than a century.

This central role of calculus was unchallenged until about five years ago, when a group of computer scientists and mathematicians argued that mathematics and its applications had changed so radically that a revolution was in order: down with calculus—up with finite mathematics! Marching orders were drawn up at a Sloan Foundation sponsored Conference held at Williamstown College in June, 1983.

In response, many colleges and universities began experimenting with the teaching of finite mathematics in place of, or in addition to, or together with calculus. Although courses in finite mathematics had been taught for over two decades, these courses had been viewed neither as foundational nor as central, but rather as additional study in one branch of mathematics. The claim now was that finite mathematics should replace calculus as the core because of the growing importance of computers and the new questions and applications of mathematics which they made possible.

My personal response to this was to think long and hard about the calculus. I eventually decided that despite the changes that had taken place in mathematics, the Williamstown conferees were wrong: calculus is as important as ever!

Calculus has stood the test of time and continues to be the foundation and wellspring of most modern mathematics. Moreover, the application of continuous mathematics continues unabated. Indeed, its applicability has been enhanced and strengthened by the advent of high-speed, large-capacity computers. In the article, "In Praise of Calculus", reprinted in this volume, Peter Lax summarizes a few recent striking developments in the mathematical understanding of dynamical systems in physics, of scattering and diffraction in wave phenomena, and of the generation and propagation of shock waves in fluid dynamics. These are just a few recent developments in calculus-based branches of mathematics that have been aided by high-speed computing.

Almost all of science is concerned with the study of systems that change, and the study of change is the very heart of the differential calculus. The description of such systems usually takes the form of an ordinary or a partial differential equation that the system satisfies. The numerical solution of such equations is one of the principal tasks of large-scale computing. Even discrete analysis of such equations using finite differences is next to meaningless without an understanding of the calculus. Thus, all science and engineering students need calculus in their studies. Moreover, it is in calculus courses that all students begin to learn the important role played by mathematics in explaining and understanding our world. Based on these facts, there is an overwhelming case for calculus remaining as the core of the undergraduate mathematics curriculum.

But, along with all of this, I realized just how much the introductory calculus course had changed, how much it had eroded in the nearly three decades since I had studied it. Although I could advance many reasons for this, I decided that I had little interest in focusing on the causes or in continuing to participate in the "calculus versus finite mathematics" debate. Rather, I felt compelled to try to improve the way calculus is taught at American colleges and universities. Although it was presumptuous to believe that I had any chance at success, I felt that I must at least try. There was simply too much at risk.

The United States is currently experiencing a shortage of young people studying mathematics, science and engineering, and this shortage is expected to worsen. Calculus is the gateway and is fundamental to all such study. Hence every student who does not complete calculus is lost to further study in science, mathematics or engineering. Moreover, many students who start calculus do not complete it successfully. The country cannot afford this now, if it ever could. Further, many of those who do finish the course, have taken a watered down, cookbook course in which all they learn are recipes, without even being taught what it is that they are cooking. Understandably, science and engineering faculties find it difficult to build on such a foundation, and they feel that they must teach their students elementary calculus as well as science or engineering. Finally, in past generations many students were sufficiently challenged and turned-on in their calculus course that they decided to become mathematicians. I don't believe that happens much today. To overcome these problems and to recapture that earlier excitement, I decided to see what could be done to improve calculus instruction.

To determine whether others shared these perceptions, Steve Maurer and I organized a panel discussion at the Joint AMS/MAA Anaheim Meeting in January, 1985, entitled "Calculus Instruction, Crucial but Ailing." Several hundred people listened to and then reinforced the panel's airing of problems connected with introductory calculus. Those assembled gave vent to their frustrations and lamented their inability to change things.

But was change really impossible? To answer this question I asked the Sloan Foundation to fund a small focused conference/workshop. My formal proposal to the Sloan Foundation follows, together with the position papers prepared for the Conference which was held at Tulane University in January, 1986. The position papers framed the beginning discussion at the Conference, and the conclusions reached are contained in the summary workshop reports.

The participants to the Conference were diverse: coming from community colleges as well as research universities, from both public and private institutions, from large and small schools, and from all parts of the country. Despite this diversity all reached one conclusion: the time was right for change. In part, dissatisfaction with teaching calculus and with the results of such teaching has grown to the point that there was unanimity at this conference that something had to be done. Further evidence of this ripeness for change is provided by the over two hundred requests that have been received for more information following the appearance of news articles on the Tulane Conference. Moreover, the group began to understand that calculus instruction is going to change! Technology is not going to let calculus instruction stay the same!

Anyone who has seen hand-held calculators which output the graph of an equation visually realizes that we can and, indeed, we will have to change what we ask students to learn and what we test them on. And this is just a start of developments that include the growing availability of programs for symbolic and algebraic methods as well as for numerical methods. As these become common on smaller machines and become easier to use they will have further and more revolutionary implications for what we should teach. These developments in technology are going to affect calculus instruction much as inexpensive calculators are affecting grade school instruction in arithmetic. Calculus instruction is going to change; the only question is whether the change will occur thoughtfully or haphazardly.

Having agreed that change is desirable, possible, and even inevitable, can we agree on the kind of change we should strive for? Surprisingly, the Conference participants did agree—participants that included an engineering dean, a physicist, and a biological statistician as well as a diverse group of faculty from the mathematical sciences. The availability of hand-held calculators such as those mentioned above and personal computers with both numerical and symbolic capabilities removes the necessity for covering many of the techniques and for much of the drill which now form such a large part of the calculus. The Conference agreed that the syllabus should be *leaner*, contain *fewer topics*, but that it should have more *conceptual depth*, numerically and geometrically. Moreover, the Conference affirmed that calculus instruction should make use of the latest technology but that the goals of the calculus must extend far beyond facility with either calculators or computers.

A sample syllabus embodying these goals was drawn up for the first two semesters of calculus, with alternatives for the second semester. These are to be found in the curriculum report in this volume. Most mathematicians will note that some of their favorite topics have been omitted or have been given only brief coverage. Although the price may seem high, this reduction in the number of topics is needed to buy the time to teach a conceptual understanding of what calculus is and why it is important. Time must also be set aside in order to experiment with giving the students more demanding problems. For not only must the content change but the testing must change, and calculus teaching must become more interactive. The students must also learn that functions are not just given by formulas but are generated by the computer or else arise from data. The centrality of the calculus in the study of systems that change must again be made clear as must the role of mathematics in modeling and understanding the real world. Teachers must have the time to excite their students about mathematics, to show them its utility and beauty, and to demonstrate what attracts people to its study.

It is well and good that agreement was achieved on the need for, and directions of, reform at the Tulane Conference. But, supposing that the rest of the community were to agree, how could we get from here to there? How do we transform current calculus instruction to instruction fulfilling the goals just described? How do we get tired faculty excited enough to devote the time and energy necessary for change? How do we get overstressed, career conscious students to be willing to work harder and accomplish more? How do we get chairs, deans, and provosts to make the improvement of calculus a priority, when their attention is often focused on cutting costs in calculus instruction or on more glamorous activities that more commonly attract their own funding? How do we get anyone to focus on an activity as well established as that of teaching calculus? In short, how can we break out of the present, familiar, though unsatisfactory situation—a conspiracy concerning calculus instruction, to which faculty, students, and administrators are parties?

The size of the enterprise is immense, and the inertia is great! As many examples attest, attempts at change, no matter how successful at first, usually revert to the present way of teaching calculus. For all of these reasons, it will take a highly coordinated, well-thought-out, national effort to effect any lasting change. But I don't believe that it is hopeless!

How might we begin? First, we should understand that we are proposing major changes in calculus instruction. Therefore, it will be necessary to create a calculus textbook written along the lines discussed above. Current texts cover too many topics, and most coverage ranges from inadequate to superficial. Moreover, even when the text is adequate, the problems are not and the emphasis is wrong. Many people argue that the main problem with calculus instruction is the current crop of textbooks. But as is argued in the article by Renz, the publishers are producing what will be used in American colleges and universities. When a nonstandard text is published, no one will use it because it is different. Therefore, before we write a new book and produce other curricular materials, we must create a market, both national and diverse, in which to test, to perfect, and eventually, to showcase the book.

Since there is no incentive for a publisher to do this, we must look elsewhere. Although leadership to plan and carry out such effort might come from one of the national organizations in the mathematical sciences, I don't believe that is likely. To be successful the calculus initiative will have to create and build a consensus for a new calculus course, both with regard to contents and methods. While national organizations such as the MAA, the AMS, or the NCTM will be extremely important and even necessary to ratify and promote the new course in calculus, it would be difficult for them to create it. Although many points of view will need to be integrated to make the effort successful, strong executive leadership will be necessary to make coherent and consistent choices. I don't believe that this will be done by a committee or by any group that must maintain consensus throughout the process. Further, this effort will require substantial resources which I doubt that any national organization would be willing or able to commit. Consequently, I believe that we will have to look elsewhere for leadership.

I envision that a reform initiative in calculus would have to be undertaken by a very small group with an executive director that is guided by a carefully chosen advisory committee. The group would have to have strong links to all parts of the mathematical sciences community. First, funding will be extremely important. Faculty and administrators will be asked to do something different, something that will take more time and effort and something that will be more expensive. Moreover, not everyone will agree at the outset on the direction taken and the choices made. The effort cannot involve volunteers because the going will be rough. Second, the source of the support will be important to give the project both stature and presence in the academic community. Therefore, it should be sought from both private and governmental foundations such as the Sloan Foundation and the National Science Foundation.

How might the project begin? To create the market for the textbook one could set up a two-year pilot program at a diverse group of colleges, both two-year and four-year, and universities, numbering about ten the first year and twice that many the second. Such a program should be closely coordinated but flexible enough to respond to local conditions. Experiences should be shared with the textbook and the other curricular materials being revised several times on the basis of actual teaching.

It should not be difficult to get experienced textbook writers interested in participating in the project. Today an author who wants to write for the calculus market is sharply constrained to follow the existing pattern. Participating in this project would enable a strong creative author to forge a new pattern, to help develop, to write for, and to test, this new course.

Although the ancillary curricular materials would have to be designed to meet the needs of the students at the different institutions that will be involved, it would seem sufficient and highly desirable that a single basic new textbook be written for the entire pilot program. Many of the benefits from participating in a coordinated national effort would be lost, and effort would be spread too thin if several different programs were being managed simultaneously. Finally, widespread adoption of the new calculus course would be much less certain if there were several new courses.

What kind of institutional support would be needed for the pilot program? First, at the schools involved there must be strong departmental commitment carried forward by a single identifiable person. This person would be both teacher and administrator for the local effort. At his institution this person must receive some released time for which the department would be compensated by external grants. Moreover, the local leader must receive a summer stipend and travel funds to participate in the planning and textbook revision workshops. The commitment on the part of the institution, the department, and the local leader would be substantial, but the fact that several institutions have already indicated a strong interest in participating based only on the news articles suggests that participants would not be difficult to find.

Finally, the last step in the initiative would be the facilitation of wide adoption of the new course through national and regional workshops. Assistance would be provided to the new group of adopting institutions during this fourth year and possibly beyond. By this point one would expect that publishers would probably be producing "nonstandard textbooks" of their own, some based on this effort and some based on other ideas on how calculus should change. The existence of a new, viable calculus course would make other experiments possible. Eventually one or more new calculus courses would supplant the present model.

As I argued above the existing calculus course will change; but whether that change is merely haphazard, or is coordinated and planned, will depend on whether or not someone takes up this challenge. It is a big challenge but much depends on it.

Ronald G. Douglas
State University of New York
Stony Brook, New York

November 1986

CALCULUS SYLLABI

Report of the Content Workshop

Jerome Goldstein   Stephen Rodi   Mac E. Van Valkenburg
Peter Lax   Thomas Tucker, Chair Paul Zorn
Louise Raphael   Robert van der Vaart

1 Introduction

This workshop was charged with developing syllabi for the first two semesters of calculus. Although very few of the conference position papers made specific recommendations for syllabi, the workshop did have CUPM recommendations from 1965, 1972, and 1981, the Advanced Placement Calculus course descriptions, and syllabi from the home institutions of many of the conference participants. This, together with the discussions of the position papers during the first two days of the conference, provided the basis for this workshop's deliberations.

Syllabi were developed for the following:

1) Calculus I, a course in the core concepts of the calculus for a general audience
2) Calculus II, a standard second semester, single variable course
3) Calculus IIM and Calculus IIC, alternative second semester courses involving either multivariable calculus (IIM) or use by students of computer symbolic manipulation programs (IIC).

The workshop spent most of its time considering Calculus I. A 35-hour, annotated course description for Calculus I is given later in great detail. Noteworthy features include emphasis on functions that are represented graphically (such as a curve on an oscilloscope or the Dow-Jones average) or numerically (from a table of data or a "black box" calculator), extensive use of hand calculators with "solve" and "integrate" keys, reduction in precalculus material and de-emphasis on limits and continuity, and a reorganization that allows treatment of all the elementary functions (including trigonometric, exponential, and logarithmic) from the first week on.

The Calculus II syllabus, although not as refined as that for Calculus I, still gives estimates for the minimum number of class hours to be spent on each topic. This course description is fairly conservative but does echo the spirit of the 1981 CUPM recommendations for second semester calculus. It should be emphasized that both Calculus I and Calculus II are "slim" courses with built-in slack time. The Calculus IIM and IIC courses were developed after the conference by Tom Tucker and Paul Zorn, respectively. Calculus IIM may be an attractive alternative to schools with a required discrete math course in the first two years. Calculus IIC may be the course of the future; the syllabus here seems to be the first such proposed nationally.

2 Boundary Conditions

Any syllabus for first year calculus must take into account a variety of considerations. What is the audience and how does it change from the first semester to the second? What do other disciplines expect from calculus as a service course? How many class-hours constitute a "semester"? What textbooks are available? What preparation can be expected of students? Are there articulation problems with high schools?

A first semester calculus class is not all engineering students, it is not all physical science majors, it is not all computer science majors, it is not all social science majors, and it is definitely not all math majors. The Calculus I syllabus is designed for a general audience, "just plain folks". If, for example, the computer science concentration requires only one semester of calculus, then it is important that that one semester cover all the "core" material from calculus. Calculus I can be taken as a terminal course—there is no postponement of integration or trigonometry or exponential functions to the second semester. It is in the second semester, where enrollments are often half that of first semester calculus, that some branching may take place. The proposed Calculus II syllabus probably best serves the engineering and physical science population, with a third semester to follow later in vector calculus. The Calculus IIM syllabus, however, may better serve the biological sciences, social sciences (economics in particular), and even computer science. If math majors are required to take a discrete math course and a linear algebra course in the first two years, then Calculus

IIM may be the only way to avoid five required courses in the first two years.

As for the number of class-hours per semester, this varies so much from school to school that the workshop decided to design syllabi that could be covered in a minimum of 35 class-hours. Schools with 40, 50, or even 60 class-hours per semester should find some slack time. Suggestions are made for uses of this slack time in each syllabus.

Questions about textbooks, student preparation, and articulation may be more matters for the Implementation Workshop, but a few comments are in order. Both Calculus I and II could conceivably be taught from existent textbooks. Van Valkenburg shows how this can be done with Ash and Ash, *The Calculus Tutorial*. This does, however, entail much jumping back and forth, closing one's eyes to some passages, and putting in additional material elsewhere. In the end, of course, it would be preferable to have a textbook written expressly for the given syllabus. One possible mechanism for the creation of such a book is a committee (NSF supported?) in the Chem Study style. As for student preparation, Calculus I assumes a knowledge of elementary functions, which means four years of high school mathematics. As Don Small's paper on articulation attests, however, many students in Calculus I have seen some calculus before in high school. The syllabi designed by this workshop do not make any special provisions for these students. Institutions where this is a problem may wish to consider some sort of accelerated program that covers three semesters of calculus in two semesters. Such a unified, two semester course with multivariable functions appearing throughout is taught at Colby College. Finally, again with respect to articulation, the impact of these syllabi should be felt immediately in calculus courses taught in high school. In particular, the findings of this conference were discussed in detail at a May meeting of College Board's Advanced Placement Calculus Committee (the chair of this workshop also chairs the AP Committee).

3 Computers and Calculators

The role of computers in calculus was discussed in some detail by the workshop. There seems to be little doubt that computer demonstrations in the classroom are desirable. The more important question is to what extent should students themselves use computers. There would appear to be three levels of usage:

1) numerical computation,
2) graphics,
3) symbolic manipulation.

The first level can be achieved on a hand-held calculator, the second on almost any micro or time-shared mainframe. At the present, the third level requires a well-equipped micro or mainframe. The obvious limitation is the accessibility of the appropriate level of hardware. It is not clear that most institutions can now support large numbers of students, say 25% of the freshman class or more, doing calculus homework on micros or mainframes. We take seriously the horror stories of waiting lines at 3:00 AM and sign-up sheets for thirty minutes on a terminal. Symbolic manipulation programs (smp's, for short) can add much to a calculus course, but it does not seem realistic to require their use at this time in mainstream calculus classes, especially in the first semester.

On the other hand, calculators are ubiquitous and can help alleviate the obsession with closed-form solutions found in most calculus courses. All of the proposed syllabi assume each student has a programmable calculator with a "solve" key (finds roots of $f(x) = 0$) and an "integrate" key (computes definite integrals numerically). There are models by Texas Instruments, Casio, and Radio Shack that have "integrate" keys and that sell for under $50. We are in contact with Texas Instruments about developing a calculator with a "solve" key also. It appears that one can be marketed at a price less than that of a calculus textbook. From our experience, it takes about half an hour for a student to learn how to program, say, the Texas Instruments 55-III. A typical function takes a couple minutes to enter into a calculator and to check that it is entered correctly; the "integrate" key takes anywhere from 15 seconds to 2 minutes depending on the number of subdivisions requested. The greatest difficulty in using programmable calculators is verifying that a given function has been entered correctly. Students will be forced to develop the number sense to recognize bad "answers" when they see them. The question of round-off error must also be addressed. Finally, unless exam questions occasionally require the use of a calculator, all of this is for naught.

The benefits are obvious. If all a student learns is how to use a calculator, how to be wary of numerical answers, how to check answers for reasonableness, it would be worth it. Nevertheless, it must be emphasized that this recommendation to use calculators should be viewed only as a stopgap. We can easily envision a micro-based, symbolic manipulation calculus course like syllabus II-C as the mainstream course in the future. The time is just not now. Instead, we make the conservative recommendation to use at least the fifteen-year-old technology of calculators that is so readily available to us.

4 Goals

Before designing syllabi, the workshop discussed the possible goals of any first year calculus course. Some overlap with the style workshop is inevitable here, even though this workshop tried to limit itself to questions of content rather than style, wherever that distinction was well-defined. Some of these goals are things students should be able to do (Maurer's "competencies" or Stein's "verbs"), some are things students should see and know (Stein's "nouns"). Among the competencies should be the ability to give a coherent mathematical argument; students must be able not only to give answers but also to justify them. Calculus should teach students how to apply mathematics in different contexts, to abstract and generalize, to analyze quantitatively and qualitatively. Students should learn to read mathematics on their own. And, of course, calculus must also teach mechanical skills, both by hand and by machine. As for things to know, students must understand the fundamental concepts of calculus: change and stasis, behavior at an instant and behavior in the average, approximation and error. Students must also know the vocabulary of calculus used to describe these concepts, and they should feel comfortable with that vocabulary when it is used in other disciplines.

In the end, students must be made aware that calculus is not taught for its own sake. The invention of the calculus in the 17th century revolutionized the way we see the world. The laws of both the natural and man-made universes come to us in the form of differential equations, equations that we are still trying to understand today, equations that can describe chaos as well as order. This past and present relevance of the calculus must be emphasized. Unfortunately, it is easy to lose sight of these lofty ideals in the daily humdrum of differentiation and integration exercises; students have other courses to take and would rather learn "how" for tests and worry about "why" later, if ever. That is why the Calculus I syllabus has "honesty" days when important concepts are first introduced. These classes should be devoted to showing why we are really interested in, say, the derivative, where it really arises. This is perhaps as much a matter of style as of content, but the workshop feels this is so important that it must be built into the syllabus.

5 Annotated Syllabus for Calculus I

The Derivative

Hours

1 *Honesty days:* The point here is to present a problem that asks for an instantaneous rate of change or local magnification constant (e.g. marginal cost, velocity). Interpret graphically as the slope of the tangent line and conclude that the desired quantity is

$$\lim_{\Delta x \to 0} \frac{f(x + \Delta x) - f(x)}{\Delta x}.$$

2-4 *A dictionary of functions:* Introduce the cast of characters:

a) x^r (r any rational number), b^x and $\log_b x$ ($b > 1$), $\sin(x)$, $\cos(x)$, and $\tan(x)$,

b) graphically presented functions,

c) numerically presented functions (from a table, a black box, a calculator),

d) new functions from old (addition, multiplication, division, composition).

What is needed for x^r, b^x, $\log_b x$ is the graph of each (consider different cases for r). Treat $\log_b x$ as inverse of b^x: $y = b^x \langle = \rangle x = \log_b y$. Do not get bogged down in properties of $\log_b$. Although we do assume audience has seen precalculus material, we probably need to give a quick review of relationship between sin, cos and (x, y)—coordinates of point on unit circle in order to emphasize periodicity. Also, remind students of radians.

For other representations, we should be honest to say that many functions (from waveforms on an oscilloscope to the Dow-Jones Average) come to us not algebraically but graphically or numerically. "Black box" keys on student calculators are a source of numerically presented functions that might be worth studying in exercises (e.g. invsinh(x)).

New-functions-from-old should include graphical interpretation of $+$, scalar multiplication, translation of origin, squaring. One kind of new-from-old construction we do not mention is splicing: we wish to emphasize the calculus problems inside the interval rather than the non-calculus problems at the end-points where we splice. Thus the absolute value function does not appear here.

This is the appropriate time to introduce students to their programmable calculators. There should be exercises requiring students to enter a given function (say $f(x) = \ln(1 + 2 \sin x)$) in their calculators, to check that the programmed function gives correct values (try $x = 0$, $x = \pi/2$), and to explore the function (is $x = 3\pi/2$ in the domain, what is the range?).

5 *Limits and continuity:* We recommend a precise english definition of "$\lim f(x) = L$" rather than the mathematically professional $\epsilon - \delta$ definition. For example, "$f(x)$ can be made arbitrarily close to L by making x sufficiently close, but not equal, to a." The less precise "small changes in x produce small changes in $f(x)$" is also useful. Numerically, a continuous function f is one whose value at any x can be computed to any given number of digits by putting into f any number near enough x. Since x is an arbitrary real number we could never actually enter x into the f-machine anyway. Graphically, a continuous function is one whose graph can be drawn "without lifting pencil from paper". Properties of limits with respect to $+$, $\cdot$, $/$ should be stated

and briefly motivated but not proved.

6   *Definition of f'(x), notation, linearity, polynomials:* Exercises computing f'(x) "the long way" from the definition of f'(x) should be few and easy (e.g., x^3, $x^2 + 5x$, NOT $1/(\sqrt{7 - 3x})$). In same hour, derive the power rule in order to be honest with class; the suspense might kill them, especially if they have seen some calculus before. Also do addition and scalar multiplication so general polynomial can be differentiated.

7   *Graphical differentiation:* Show how f'(x) can be graphed given only graph for f(x). This will automatically lead to relationship between qualitative behavior of f(x) and sign of f'(x).

8   sin(x), cos(x), b^x *and the number e:* Graphical differentiation should motivate $(d/dx)\sin x = \cos x$. Actual proof can be downplayed, especially the proof that $\lim_{h \to 0} \sin h/h = 1$. For $f(x) = b^x$, quickly derive that $f'(x) = kb^x$ where k is the slope at $x = 0$. Then define e to be the number such that slope at $x = 0$ is 1 (slope at $x = 0$ certainly looks like a continuous function of the parameter b—"small changes in b produce small changes in slope at $x = 0$").

9   *New-from-old differentiation:* Derivative of sums, products, reciprocals, and quotients. Since the negative power rule follows from the quotient rule, we might as well admit that the power rule holds for all rational exponents so that we can assign some interesting exercises. Actual proof must wait until hour 10 or 11.

10 *The chain rule, inverse functions, ln x, $\sqrt{x}$:* Give chain rule in both dy/dx and $f'(x)$ notations. Use chain rule to show that if $y = g^{-1}(x)$, then $dy/dx = 1/g'(y)$ and apply to log x and $\sqrt{x}$.

11 *Implicit differentiation:* Treat as a formal, manipulative process motivated by chain rule notation: $d/dx(y^3) = 3y^2 \, dy/dx$. Avoid theoretical considerations of implicitly defined functions (implicit function theorem and partial derivatives).

Uses of the Derivative

12-15   *Qualitative analysis of functions (curve sketching):* This should include the following:

   a) sign of f', increasing/decreasing, critical points;
   b) sign of f'', concavity, points of inflection;
   c) max/min on a closed interval, candidates;
   d) graphical relationship of f' and f (given the graph of f', find the graph of f).

Use all functions in dictionary, not just polynomials. Include f(x) with messy f'(x) so "solve" key on calculators can be used. Distinguish rela-

tive (local) max/min from absolute (global) max/min. Note that latter problem involves the non-calculus end-points as candidates and also becomes easy with calculators. Do not emphasize max/min at points where derivative does not exist.

16-17   *Root-finding, Newton's Method, Rolle's Theorem:* Mention the non-calculus bisection or secant methods as well as the calculus Newton's method. Explain what the "solve" key is doing. Question of finding all roots, not just one, should arise naturally. That is a good time to bring in Rolle's theorem (two roots of f(x) must have a root of f'(x) in between), and hence the Mean Value Theorem.

18-19   *Linear approximation and error, big "Oh" notation:* Derive the equation:

$$f(x) - f(a) = f'(a)(x - a) + f''(c)(x - a)^2/2,$$

and interpret with big Oh notation:

$$\Delta y = f'(a)\Delta x + O((\Delta x)^2).$$

That the error in the tangent line approximation is $O(\Delta x)^2$) should be observable on students' own calculators. To show that approximation is still important even with calculators, there should be exercises on relative (percentage) accuracy in measurement. Big Oh notation should be welcomed by computer scientists. To avoid confusion, little oh is not introduced.

20-21   *Extrema ("word") problems:* Maximizing volumes of boxes and silos (does Farmer Brown really use calculus to fence in his cows?) is OK for teaching problem-solving strategy, but it might also be nice to see, for example, Snell's Law.

The Integral

22   *Honesty-day:* Again, honesty means telling students what we are going to do and why. Summing examples include work, present value of money, total distance, any averaging problem (temperature, depth of river), as well as area. Contrast should be made with derivative: total versus instantaneous, global versus local information, (formally) functional versus operator. Should finish with definition of $\int_a^b f$ as the limit of Riemann sums.

23-24   *Numerical integration—rectangle, midpoint, trapezoidal, and Simpson's rule:* Discussion should include enough error analysis to give, without proof, the order of the error (big Oh again). Notice that rectangle rules (left, right, upper, lower) and midpoint rules are valid

Riemann sums. Trapezoid rule and Simpson's rule are not. A nice way to view trapezoid is weighted average of left and right rectangle (or weighted average of function values with half weights at end points because "half" of their influence lies outside interval). It is easy to see with picture that trap and midpoint have errors of opposite sign and that both errors depend on concavity (i.e. f'') and that midpoint's error is half of trap's for a parabola. This suggests a weighted average of twice midpoint and once trapezoid. This is Simpson's rule. Students should be told that the $\int$ key on their calculator, which they should begin using, might be based on Simpson's rule. Again, students should be able to observe the order of the error on their calculators. They should begin to learn that calculators do not give the "exact" answer, if they haven't learned that already. They also will need to be able to give rough estimates for the value of a definite integral in order to check that calculator answers are reasonable.

25   *Properties of the definite integral:* The properties needed are:

a) linearity
b) $\int_a^b = \int_a^c + \int_c^b$
c) $m(b - a) \le \int_a^b \le M(b - a)$.

Change of variable in integrals, $\int_a^b f(g(x))g'(x)dx = \int_c^d f(y)dy$, can be justified by Riemann sums and the Mean Value Theorem.

26–27   *The area function,* $A(x) = \int_a^x f$: This is another example of new-from-old, but students distrust it because $A(x)$ appears to be not computable and hence abstract. With calculators having an "Integrate" key, the abstract $A(x)$ can become concrete. Of course, $A(x)$ has always been concrete graphically and there must be exercises sketching $A(x)$ given the graph of $f(x)$ (do not use that $A(x)$ is an antiderivative of $f(x)$—let that become evident). The actual use of $A(x)$ as *c.d.f.* in probability could be mentioned and an interpretation of $A(x)$ in terms of an application from "honesty" day should be given. Finally, $A(x)$ can be computed explicitly for $f(x) = mx + b$.

28   *The Fundamental Theorem of Calculus: After 25-27 the whole class should scream* $A'(x) = f(x)$. *The FTC in the form* $\int_a^b f = F(b) - F(a)$ *needs to use* $F'(x) = A'(x) \Rightarrow F(x) = A(x) + C$. *Either treat this "obvious" fact intuitively or prove using Mean Value Theorem.*

29–30   *Antidifferentiation:* Do $\sin x$, $\cos x$, e^x, $1/x$, $\ln x$ (given from oracle as $x \ln x - x + C$ but easily checked). Students should also be able to antidifferentiate $x \sin(x^2)$ by substitution or guess-and-check.

Uses of Integral

31–32   *Area between curves and average value:* Might go back and reinterpret FTC in terms of the average value of a function, if integral has not been viewed as an average from the beginning. Applications might also include arc length to show off the "integral" key. Again, with calculators, area between arbitrary polynomials is feasible. Consider also work, present value as well as volume.

33–35   *Easy differential equations:* Topics should include:

a) acceleration-velocity-position problems
b) exponential growth: $y' \pm ky$
c) cyclic behavior: $y'' = -y$.

For (a) there should be examples of nonconstant acceleration, in order to avoid the formulas students have memorized from physics. Note for (b) that the growth rate of b^x compared to x'' does not need l'Hôpital's rule: just ask students to compare 2^{100} with 100^2 or 100^3 or 100^{19}. Both (b) and (c) are good places to explain the role of the calculus in describing the natural and man-made universe. Kenelly's paper is particularly appropriate for (c).

General Comments. It is worth noting what is not in this course: related rates, l'Hôpital's rule, $\epsilon - \delta$ definition of limit, and precalculus material on lines, circles, and domain and range of functions. Other material is greatly reduced: limits and continuity, computing derivatives from the quotient definition, computing integrals from the Riemann sum definition (using formulas for $\Sigma_{i=1}^m i^2$ etc.). There are, of course, also topics in this syllabus not usually covered in a standard first semester calculus course: graphical and numerical treatment of functions, big Oh notation, Newton's method, numerical integration, the use of hand calculators.

The allotment of 35 hours is a minimum. If more time is available, it is not a good idea to try to cover additional topics. If n hours are available, then a topic given x hours in the syllabus should be given $xn/35$ hours. That may be a glib response to the question of slack time, but it does reflect the workshop's feelings about content-crammed courses. The style workshop has many suggestions for calculus pedagogy, some of them quite time consuming. Slack time in Calculus I is the perfect opportunity to try out some of those suggestions.

6 Syllabus for Calculus II

Hours

1–2   *Expanded dictionary of functions:* tan, sec, arcsin, arctan, sinh, cosh. Give enough to read a table of integrals—definition, graphs, elementary identities $(1 + \tan^2 = \sec^2, \cosh^2 - \sinh^2 = 1)$ and formulas for derivatives.

3-5   *Techniques of antidifferentiation:* substitution, easy integration by parts, tables of integrals.

6-7   *Improper integrals:*

$$\int_a^\infty \frac{1}{x^2}, \qquad \int_0^a \frac{1}{\sqrt{x}}, \quad \text{NOT} \quad \int_{-1}^1 \frac{1}{x^2}$$

8-9   *Parametric equations:* elimination of parameter, relation between curve and dx/dt, dy/dt.

10-12   *Sequences:* include discussion of recurrence relations with examples from population models and iterative numerical procedures: limit of a sequence.

13   *Honesty day:* why study series?

14   *Convergence:* concepts yes, tests no. Examples should include geometric, harmonic, and alternating series.

15-17   *Taylor series with error form:* the idea of higher order approximation, radius of convergence from ratio test.

18-19   *Dictionary of Taylor series and manipulation of series.*

20   Order of magnitude (big Oh) and Taylor series (to replace L'Hôpital's rule).

21-23   *Complex numbers and polar coordinates:* do $e^{i\theta} = \cos\theta + i\sin\theta$ with series and geometrically.

24   *Honesty day:* where do differential equations come from?

25-27   *Graphical methods:* vector field and phase plane and numerical analysis with models.

28-30   *Closed form solutions:* variables separable, constant coefficient linear using complex numbers for order $n > 1$.

General Comments. This course represents a core of material for the second semester. Some topics, such as exponential/logarithmic functions or applications of the definite integral, are not here because they are already in Calculus I. Other topics are simply not mentioned at all: logarithmic differentiation (e.g., $x^{\sin x}$), indeterminate forms such as $\infty - \infty$, special techniques of integration including trigonometric substitution and partial fractions, convergence tests for series of numbers, integrating factors for linear differential equations. Material that might be considered new or unusual includes: recurrence relations, big Oh analysis with Taylor series, complex numbers, the emphasis on graphical and numerical analysis of differential equations. The de-emphasis of convergence tests means this course does not look forward to a later majors course in analysis as much as most present Calculus II courses do.

This course is even slimmer than Calculus I. Again, the general advice is to spend the slack time teaching better

rather than more—try out some ideas from the style workshop, keep using hand calculators, do more modeling. On the other hand, the slack time may be sufficient to allow one extra favorite topic; for example, the workshop seemed reluctant to part with partial fractions.

7 Syllabus for Calculus II—Computer Alternative

Hours

1,2   *Honesty days:* Why integrate? When in closed form? When numerically? Advantages of each. Examples of everything. Review Fundamental Theorem: it says two ostensibly different things are really the same. Provisional estimates of standard definite integrals: $\exp(-x^2)$, $\cos(x^2)$. Exercises: some numerical integrals by hand, no machines; graphical integrals; compare with closed form solutions.

3,4,5   *Numerical integration, revisited:* (Some methods already seen in semester I. Here emphasize error analysis.) Trapezoid, midpoint, Simpson's rules, how to use smp to compute them. Comparison of observed errors for each method. Error formula for each method: don't prove, but use them in realistic ways. Use smp to compute or estimate graphically extrema of f'' and $f^{(4)}$. Maybe an application, e.g. to arclength where closed-form techniques usually fail. Exercises: estimate $\int_0^1 \exp(x^2)dx$ with three decimal place accuracy; compare error estimate with observed error for familiar integrands.

6-9   *Techniques of antidifferentiation:* as in main II syllabus, except that partial fractions are included—smp can be used to expand rational functions in partial fraction form; the point is that algebraic manipulations transform integrand to more tractable form. Important theme for this unit: conjectured answers can always be checked by differentiation, so do anything necessary to generate reasonable conjectures. Thus, e.g., include method of undetermined coefficients for certain trigonometric integrals. When using smp as oracle, say so clearly.

10,11   *Improper integrals:* Use smp to illustrate numerically how integrals can either converge or diverge. Some error analysis: how tails of, e.g., $\int_{-\infty}^\infty 1/(x^2 + 1)dx$ get small. Application to normal probability distribution: where does the Z-score table come from?

12,13   *Parametric equations:* Use smp for graph-sketching, and for computing arclengths (by numerical integration) for interesting curves, like cycloids, hypocycloids.

14-16   *Sequences:* Rationale as in II syllabus; smp's invaluable here, to list numerical terms of sequences, including recursively defined. Newton's and bisection method iterates as examples. Comparison of speed of convergence. Smp's allow some exercises on understanding definition of convergence. e.g., given a sequence and its limit, how many terms until deviation from limit less than .1, .01, .001, etc.)

17   *Honesty:* Why bother with series? How does calculator compute $\sin(1)$? Where does polynomial estimate $p(x) = x - x^3/6 + x^5/120$ to $\sin(x)$ come from? How do graphs compare? Is derivative of $\sin(x)$ approximated by derivative of p? Use smp to raise, investigate such questions.

18,19   *Convergence of series of numbers:* Standard examples. To illustrate concept, smp generates numerical examples. (Example command: $\text{Sum}[n \uparrow (-2), \{n, 1, 20\}]$ adds 20 terms.) Geometric and alternating series, studies because partial sums can be either computed in closed form or estimated easily. Smp's used, e.g., to estimate $\sin(2)$ to many decimals by alternating series. *P*-series studied as important examples, and to draw analogy between series and integral. No formal convergence testing.

20–25   *Taylor's theorem with error formula:* Idea of estimating functions by series is main motivation for studying series. Include big Oh analysis of Taylor polynomial error. Smp's can compute and graph the higher derivatives involved in the Taylor error formula, as well as compute numerical estimates to values of standard transcendental functions, and to integrate such functions by series method. (Total time on this topic reduced by one day from II syllabus: with smp to expand arbitrary functions in series, less need for series dictionary.)

26–28   *Alternate plan—Multiple integrals:* Smp computes double Riemann sums for integrals over rectangles, draws surfaces. (Students often misunderstand double Riemann sums; smp may help show numerically how they approximate integral, just as in one-variable case.)

29   *Differential equations honesty day:* Why bother? What do they do? Good modeling examples: how do real problems lead to DE's, whether or not we can solve them? Closed form vs. numerical methods? Which are easy and which are hard? Graphical techniques: smp's can draw direction fields; point is that solutions exist, can be estimated, even if not available in closed form.

30–32   *Closed-form techniques:* Separable DE and first-order linear. Physical models: free fall with air resistance, mixing, population. (Omit higher-order DE's unless complex arithmetic was covered above.)

33–35   *Numerical DE:* Euler and Runge-Kutta methods. Smp's carry out most of the tedious manipulation. Compare observed errors in simple cases; don't treat error formulas rigorously. Series methods (a nice place to see recursive formulas for coefficients of the solution.)

General Comments.   The abbreviation "smp" is used to stand for any powerful computer algebra system that has graphics. It should be understood that smp's don't require programming by the user (although they do allow it; some students will discover this and make good use of it.) Think of an smp as a super calculator, that does for symbolic, numerical, and graphical operations what calculators do for numbers. Smp's are really almost that easy to use. They are also much more convenient than calculators because they remember their work. Thus, for example, one defines a function only once, and refers to it by name thereafter, when integrating, differentiating, graphing, expanding in a series, etc. Experience shows that one day devoted to the mechanics of the smp itself is enough: students pick up other commands and a general feeling for the system as they go along.

Having smp's available allows one to treat approximate and numerical methods more realistically than would otherwise be possible. For example, error formulas may involve extrema of high derivatives of complicated functions. These are hard to compute or even estimate by hand, but with an smp, one can just calculate the derivative symbolically and find its extremum either graphically (probably good enough) or by setting the next derivative to zero.

The main advantage of teaching some numerical and graphical methods is to strike a balance between them and closed-form techniques. The latter are usually overemphasized. The point is that calculus offers a variety of techniques—one chooses what is appropriate to the given situation. A recurring theme of this course should be that the methods complement each other. Hence the course is structured to present something of each approach to the various topics raised: integration, series approximation, and differential equations. One advantage is simply to make the point that most of the problems raised in calculus have answers, even if a particular method doesn't readily produce them (e.g., the definite integral of a reasonable function exists, even if the fundamental theorem doesn't compute it—many students seem to believe that the fundamental theorem *defines* the definite integral.)

The different smp's available differ in how easily and successfully they handle antidifferentiation. Macsyma is excellent; SMP is fairly awful. This shouldn't pose a serious problem. The push-the-button-and-get-the-answer aspect of smp's should be played down anyway at this level. Better to use the smp to do things that one "could" do anyway, given enough time and patience and to convince students that this is so. (One way: on tests, and in exercises, sometimes insist that everything be done by pencil—say, compute the first three Newton's method iterations to the square root of 6 as rational numbers. Bring the stone age to the silicon age.) Almost all smp uses proposed above are of that kind.

The syllabus suggested runs to 35 days, like the first semester syllabus. Another difference from the calculus II syllabus is that some baby two-variable integration (supported by smp computation of double Riemann sums) is given as an optional replacement for the three days Calculus II devotes to polar coordinates and complex numbers. One is loath to jettison the latter topics, but no use is made elsewhere of polar coordinates, and complex numbers are used only in higher-order linear DE, which I recommend omitting also. (At St. Olaf, and perhaps elsewhere, client departments want multiple integrals in the first year.)

In many parts of the syllabus, exercises are suggested to use smp's. These give the flavor, but don't exhaust the possibilities.

8 Syllabus for Calculus II—Multivariable Alternative

Hours

1-2   *Expanded dictionary of functions:* same as Calculus II syllabus.

3-5   *Techniques of antidifferentiation:* same as Calculus II syllabus.

6   *Improper integrals*

7   *Honesty day:* where do multivariable functions come from? Obvious examples come from physics but economics is also a good source. A glimpse of the complications caused by the more interesting geometry of a two-dimensional domain should be given.

8-10 *Two and three dimensional analytic geometry:* vectors, dot product, cross product, lines and planes.

11-12 *Alternative coordinates:* polar, cylindrical, and spherical coordinates. Application: 3-dimensional computer graphics.

13-14 *Curves:* Parametric equations, arc length, velocity and acceleration vectors. Possible application: derivation of Kepler's laws from Newton's laws.

15-16 *Partial derivatives and chain rules*

17-19 *The gradient:* level sets, tangent planes to explicitly and implicitly defined functions, directional derivatives.

20   *Linear approximation and the differential*

21-23 *Critical points and local behavior:* Maxima, minima, saddle points illustrated with level curves and computer graphics.

24   *Lagrange multipliers and constrained extremal problems*

25-28 *Double and triple integrals in rectangular coordinates:* include some discussion of numerical techniques.

29-30 *Change of variable in integrals:* polar, cylindrical, and spherical coordinates.

31-35 *Elementary vector fields:* line integrals and work, conservative fields, surface integrals and flux, Green's theorem.

General Comments. Clearly most of the standard Calculus II material is missing, especially series and differential equations. Many mathematics departments may wish to require for all majors a later sophomore or junior level course in series and differential equations. In fact, except for the six hour prelude of single variable calculus, this is really a middle-of-the-road multivariable course that could be taught out of any thick calculus textbook.

We have not made an attempt to rethink the treatment of the topics in the manner of Calculus I. On the other hand, the spirit of Calculus I and the pedagogical suggestions of the Style Workshop can still be applied. In particular, we still urge the use of calculators.

Only a few applications are mentioned in the syllabus. The derivation of Kepler's laws should be familiar. The computer graphics application in hour 12 refers to the problem of computing the two dimensional screen coordinates of a point given the rectangular coordinates of the viewer's location. Not only does this make use of lines, planes, projection, and spherical coordinates, it also explains what the computer does to create the pictures of surfaces needed later in hours 21–23.

Finally, although much of the motivation for multivariable calculus comes from physics, there are also some nice applications from economics, such as Edgewater boxes, Pareto optimality, and level sets of utility functions.

NOTES ON TEACHING CALCULUS

Report of the Methods Workshop

Robert B. Davis	Katherine P. Layton	Sherman K. Stein
Susanna S. Epp	Alan H. Schoenfeld, Chair	Steven S. Terry
John W. Kenelly	Lynn A. Steen	H. R. van der Vaart

1 A Rationale for Change

For many years calculus has held a special place in the college mathematics curriculum. It has been *the* introductory mathematics course, and for good reason. Properly taught calculus courses served a variety of audiences, and served them well. They served as an introduction to "what mathematics is all about" for liberal arts students, as an introduction to the "language of science" for those who would go on to use mathematics, and as an introduction to fundamental mathematical notions for those who would go on to be mathematics majors. Calculus was, and deserved its role as, a foundation for college mathematics.

As we are all too painfully aware, there are now large cracks in that foundation. There appears to be general dissatisfaction with the calculus, both among students and faculty.[1] This dissatisfaction does not seem to arise simply from the availability of alternate first year courses such as discrete mathematics, although the presence of such courses may suggest that "calculus for everyone, no matter what" may no longer be appropriate. Rather, it comes from the perception that calculus courses, as currently taught, do not meet the needs they were designed to meet—and once did meet. Among the complaints frequently voiced about the state of the art, the following are frequently heard:

- that current practice is perceived to drive students away from scientific and mathematical careers.
- that most current courses are superficial, and that typical "mimicry calculus" courses do not develop understanding. They fail to prepare science students for applications of mathematics to their disciplines, and fail to convey to mathematics students a sense of mathematics and mathematical thinking.
- that the range of options in calculus as currently taught (as defined, for example, by standard textbooks) is far too narrow; that the scope of mathematical thinking, the expectations of students, and the range of behaviors expected of students are all too narrow.
- that calculus is a "stepchild" in many departments, a course with large captive enrollments receiving minimal attention from tenured faculty.
- that the departments for which calculus is a service course are increasingly unhappy with the course. For example, the IEEE has just taken the extraordinary step of publishing its own calculus book. Mathematics departments run the risk of (once again!) having their service constituencies offering their own calculus courses.
- that calculus as presently taught is essentially irrelevant for the nearly 50% of the college students who do not go on to use mathematical tools in their careers; that these students, who were once well served by a "liberal arts" introduction to mathematical thinking via calculus, are ill-served by current versions of the course.

Our purpose in listing these complaints is not to bemoan the current state, but to indicate that there is cause for concern, and reason to change. It is our belief that properly designed and taught calculus courses can and should meet the needs of the student groups mentioned above (liberal arts students, science majors, and mathematics majors). We believe it is essential to teach such courses. We believe that some significant improvements in instruction can be achieved without additional allocations of resources, and urge faculty and departments to

---

[1]There are, of course, many exceptions. At many campuses where calculus instruction receives high priority, students and faculty are quite content with course offerings.

begin making such improvements as soon as possible. We note that some of the changes we recommend will call for allocating more resources to calculus instruction than is current practice. We urge departments to allocate resources to calculus instruction where possible, and to seek additional resources from university administrations when such resources are necessary.

The balance of this report addresses a variety of issues relevant to the "delivery" end of calculus—how we can teach the subject matter in a way most beneficial for our students.

2 Goals for Calculus Instruction

Calculus is the language of change. It is a domain of rich and powerful ideas—rich in terms of intrinsic interest and their demonstration of fundamental mathematical notions, and powerful in the scope of their applications. It is the first college mathematics course for many students. It is also the last exposure to formal mathematics for many of those students. For this reason, among others, it is essential that the course help students learn to "think mathematical"—that is, to use mathematical tools as a means for solving problems, and mathematical ideas for making sense of complicated situations. With these general comments in mind, we describe the following goals for calculus instruction. We note that the goals are not in priority order. A goal should not be considered less important because it appears near the bottom of the list.

Goals for Instruction in Calculus

- Calculus instruction should develop students' understanding of concepts, as well as their ability to use the relevant procedures, in a select set of fundamental calculus topics.[2] Instruction should be aimed at conceptual understanding, and at developing in students the ability to apply the subject matter they have studied with flexibility and resourcefulness.

- Calculus instruction should expose students to a broad range of problems and problem situations (ranging from exercises to open-ended problems and exploratory situations), and a broad range of approaches and techniques for dealing with them (ranging from the straightforward application of the appropriate formulas to the use of approximation methods and modeling techniques).

- Calculus instruction should help students develop an appreciation of what mathematics is, and how it is used.

- Calculus instruction should help students develop precision in both written and oral presentation.

- Calculus instruction should help students develop their analytical skills, and the ability to reason in extended chains of argument.

- Calculus instruction should help students develop the ability to read and use text and other mathematical materials.

3 What it Takes to Do the Job Right

There was a consensus among conference participants that the learning of mathematics, especially at introductory levels such as calculus, requires frequent feedback. It was agreed that there is no one particular structure that is "right" for such instruction. Thus the issue of teaching resources is not appropriately framed, for example, in terms of the "small classes vs. large" controversy. Rather, the issue is whether any particular instructional format provides the kind of experiences that are necessary for students to master mathematics at more than a purely mechanical level.

Mathematics has been called the language of science, and an analogy to language learning strikes us as particularly appropriate. One of the goals for the calculus course is that students develop precision in written and oral work. In order for students to develop the ability to write mathematics correctly, students require frequent and detailed comments on at least some of their written work. As in English composition classes, some of the reading needs to be done by the faculty (or at least by advanced graduate students). As in composition classes, this is a laborious and time-consuming procedure. If students are to learn to "speak" mathematics, they will need opportunities similar to those in language classes, or in language laboratories. That is, they need opportunities to present their work orally in class, and to have their presentations critiqued. Providing good feedback in a calculus course is as essential for correcting students' mathematical misconceptions as, for example, providing good feedback in tennis is essential for correcting a bad serve. In the absence of such coaching, students develop misunderstandings that are hard to unlearn, and that become obstacles to further learning.

We note that in many departments substantial improvements can be made by assigning higher priority to calculus and by making the appropriate internal resource allocations (including the time, energy, and rewards devoted to calculus teaching). We also note, however, that current practice often makes it unduly difficult to teach calculus effectively. For example, difficulties are often caused when students enter a course with widely varying levels of preparation. (The first few weeks of many courses are spent in review, to give weaker students a chance to catch up. Many of them don't.) The course out-

---

[2]In order that this goal be achieved, the calculus "core" needs to consist of fewer topics than are currently crammed into the curriculum. See the report from the "content" workshop for a suggested syllabus.

lined in the content panel's report is a "no-nonsense" course that begins with real calculus on the first day. It is essential that students be ready for it—and it is thus essential to have an effective diagnostic and placement program. Many of the current difficulties with calculus would be lessened if only those students who are prepared to take courses are permitted to enroll in them.[3] It is important to have adequately trained teaching assistants, for their role in such courses (at least for many course structures at many institutions) is significant. As noted in the next section, there are a number of low-cost TA training programs that departments can adopt.

It may be appropriate for departments to explore a variety of course structures to determine the ones that work for them. As noted above, composition courses are generally taught in small sections with frequent and detailed feedback. Elementary language courses have language labs, and elementary science courses often have large lectures, but smaller laboratory sections where students receive more individualized training. Comparable structures may be appropriate for some mathematics courses.

Unfortunately, many mathematics departments do not have the resources to teach the kinds of calculus courses described in the previous paragraph. As calculus enrollments have grown, mathematics departments have first cut back on feedback and then on small sections. The results of such cutbacks were described in Section 1 of this report. If we take seriously the idea that mathematics is the language of science, and we expect students to learn to read and write that language, then resources comparable to the resources currently allocated for language instruction are essential to get the job done. In general mathematicians have been quite naive about university politics, and the issue here is a political one. If colleges and universities wish to see mathematics instruction meet the goals described in this report, they will need to allocate the appropriate resources to the task. Mathematicians will have to make this clear, and they will have to argue convincingly for the needed resources.

Though the focus in this section has been on allocation of time and energy, it is clear that a different kind of resource would be tremendously useful as well. The community would profit greatly from information and materials that would assist in the delivery of good calculus instruction. Some exist, and are mentioned in the balance of this report. We also hope to see specific kinds of materials developed, both for use with students and to assist the faculty. See section 9 for details.

4 Some Suggestions for Teaching

Achieving the goals outlined in Section 2 may call for employing a variety of teaching methods. As is always the case, instructors will feel comfortable with some methods and not with others. Our intention here is to point to some possibilities that the reader may find worth considering.

Three MAA publications are helpful in this regard.[4] *College Mathematics: Suggestions on How to Teach It* provides a basic description of useful classroom techniques, although it is mostly lecture-oriented. *Training Programs for Teaching Assistants in Mathematics* describes some low-cost TA Training Programs. MAA Notes #1, *Problem Solving in the Mathematics Curriculum*, contains fifty pages of suggestions for teaching problem solving. Many, if not all, of those suggestions can be adapted for use in standard instruction.

During the conference that produced this report, numerous suggestions were made regarding techniques for teaching calculus that individuals had used successfully. Other suggestions appear in the papers that comprise the conference proceedings. The following techniques were recommended:

- the use of complex problems from the "real world" to serve as a context for doing mathematics, and for introducing mathematical ideas (See Bob Davis' paper).
- the use of elementary theoretical problems as "cognitive bridges" to help students develop their understanding of theoretical ideas (See Susanna Epp's paper).
- the use of occasional non-standard, context-free problems (See Sherman Stein's paper).
- replacing many "show that" problems with equivalent "is it true; provide a proof or give a counterexample" problems (and, of course, assigning some similar problems for which a plausible conjecture is not true). In general, having students construct examples.
- assigning multi-step problems, and other problems that go beyond the "plug into the technique we just studied" mode.
- assigning some problems that may take two or three weeks to solve, and allowing students to hand in preliminary attempts for comment.
- assigning a collection of problems at the beginning of the term, which can be solved at various points during the term (and not revealing when they are solvable).

---

[3]Failure rates in many courses are reported at 50% and above. We believe that with a good placement program, with high quality instruction, and a good support structure for a course, failure rates should be below 15%. A failure rate above 15% indicates that there are problems with either placement, instruction, or support structure.

[4]Information about these and other MAA publications can be obtained by writing the Mathematical Association of America, 1529 Eighteenth Street N.W., Washington, DC, 20036.

- giving problems to be worked in class, by groups of three or four students.
- giving mathematics reading assignments, where students are instructed to read a text or other specially prepared materials, and then to work problems based on those readings.
- giving assignments to be worked by groups of students rather than by individuals.
- giving assignments that call for coherent written arguments, and grading them according to quality of exposition.
- having students give "formal presentations" in class or in specially arranged sessions.
- giving oral examinations.
- using graphical or tabular representations of functions as well as algebraic ones.
- using a variety of techniques for exploiting technology; see section 5 of this report.

We hope that the MAA and other groups will compile more extensive sets of suggestions (fleshed out with typical examples, classroom protocols, etc.), and that they will work to make useful classroom materials available.

It is important for instructors to define their expectations for class performance clearly, and very early in the term. A course that emphasizes understanding may be novel to the student, who will need to understand the "rules of the game." It is also important to give some typical assignments, graded according to the appropriate standards, early in the term. If new behavior patterns are established early, there is a good chance they will take root. If, however, students can proceed through the course for the first few weeks in a relatively mechanical way, it may be too late at that point to make the appropriate changes.

Especially because students may not be used to the expectations or demands of courses that stress understanding, it is important for the instructors' presentations to reflect the kinds of thinking that students are expected to develop. Thus, for example, when presenting a theorem, it is useful to do more than "motivate" the theorem, state it, and prove it; if time permits (and time does permit in the revised syllabus suggested by the content panel), theorems should arise as the "natural" answers to interesting questions. The rationales for the proofs, if not the proofs themselves, can be worked out as problem-solving tasks by teacher and students together. Also, it was observed at the conference that students' class notes are, more often than not, verbatim copies of what the teacher has written on the blackboard. Students will take those notes as their models of the appropriate type of mathematical exposition. Hence it is important that our own boardwork meet the appropriate standards of coherence and completeness.

Finally, we should comment on the core calculus curriculum itself. The content workshop members deliberately selected a "slim" curriculum. Their suggested first semester syllabus can be covered in 35 hours, and the second in 30. Often one has much more time available. If you do, we urge you to resist the temptation to add more material under the assumption that "more is better." More is not better in this case. It is more important for students to understand a small number of fundamental ideas than to deal with a large number of topics superficially. We recommend that you not add more topics to the syllabus, but that you use the time you have to explore the fundamental ideas of calculus in greater depth.

5 Dealing with Technology in Calculus Classes

Given the central role that calculus plays in preparing students for careers in science, it is important that the course (where appropriate) introduce such students to the mathematical tools that they will be using. Given that calculus is, *de facto*, the course in "mathematical literacy" for students who will not pursue scientific careers, it is important that the course (again, where appropriate) illustrate the use of contemporary technology as it applies to mathematics.[5]

We encourage the appropriate use of relevant technologies (in particular, the use of calculators and computers) in calculus instruction. It is important for students to be introduced to contemporary tools for mathematical analysis and students will profit from using calculators and computers. Many topics are dramatically illustrated with the help of technology. Moreover, the use of technological tools to do computational hackwork can free both teacher and students from computational tedium—thus allowing them to focus on conceptual rather than computational matters.

Many calculator and computer-based technologies are available for classroom use. Indeed, one recommendation of the content panel at the calculus conference was that every student be required to obtain a programmable calculator with a "solve" key and a numerical integration key, and that the calculus course take advantage of the power of such tools. Since the content panel's recommendations address the uses of hand-held calculators in calculus, we shall focus on computer-based technologies.

---

[5]Although our focus is primarily on electronic technologies, we should note that a variety of technologies are useful in (and rarely found in) mathematics classrooms. For example, prepared graphs or multiple overlays on overhead projectors can often illustrate points much more clearly than sketches at the blackboard. Xeroxed class notes can free students from the need to copy long arguments into their notebooks and give the teacher more time to discuss main ideas. A range of films, including some produced by the MAA, illustrate mathematical points quite nicely. Videotapes can be used as parts of TA training programs, or to allow faculty to review their teaching performance. Videodiscs are a new technology, and one that may prove useful.

Computer-based tools for potential use in calculus include number crunchers and graph crunchers, computer algebra systems, calculus teaching kits, computer-assisted instruction and intelligent tutoring systems, and microworlds.

Number crunchers enable a class to deal with "real" problems and illustrative examples from the sciences. Graphing programs can relieve the tedium of sketching simple curves. They can allow one to use more complex functions for analysis than otherwise possible. They can also be used to get across a number of theoretical points—for example, showing how well (or poorly) the sum of several terms in the Taylor series approximation to $f(x)$ converges to $f(x)$, or demonstrating the asymptotic behavior of functions that have been analyzed symbolically.

Computer algebra systems (among them mu-math, MACSIMA, TK-Solver) now perform tasks that are difficult to perform by hand. Complicated integrals can be evaluated, symbolically or numerically; nasty equations can be solved, etc. Specific examples of the use symbol manipulation packages in calculus may be found in the conference papers by Paul Zorn and Don Small.

There are now on the market a number of software packages designed for calculus instruction. It is not our purpose here to evaluate them, but to note their existence. There are also a number of packages for computer-assisted instruction (especially for algebra and trig, which may be used for remedial work), and "intelligent tutoring systems." These systems, much more sophisticated than the old "drill and practice" programs, build models of the student's performance and select sequences of instruction and practice problems based on detailed assessments of the students' work.[6] Though such programs do not exist for calculus at present, it is reasonable to expect them to appear in the near future—and it is reasonable for the mathematical community to try to take advantage of them. The same comment applies to "microworlds," self-contained computer-based environments such as LOGO. LOGO has been used for simulations of force interactions at the level of college physics; we hope the community will try to exploit similar environments for college mathematics.

This report can only suggest the wide range of applications of technology to the teaching of calculus. Fortunately, the MAA's Committee on Computers in Mathematics Education is currently preparing a report on the use of computers in mathematics instruction at the college level. We hope that more reports will become available, and that other materials (sample protocols of successful classroom sessions, illustrative examples of the use of technology, etc.) will be produced.

---

[6]See, e.g. Derek Sleeman and John Seely Brown's *Intelligent tutoring Systems* (Academic Press, 1982).

6 Notes on Testing

It is an old (but true) saying that "tests drive the curriculum." Examinations give the students the real "bottom line"—what they are *really* expected to know, whatever the classroom rhetoric about "understanding" may be. That bottom line, of course, determines what students will study—and learn.

Standard hour-long examinations serve a valuable purpose. They are limited, however, in that they examine only a subset of the students' skills. Section 2 of this report listed a broad set of goals for calculus instruction. If we take those goals seriously, we will need means of evaluating students' progress towards them—both because testing those competencies will make sure that students work to develop them, and because we need to know ourselves whether our instruction is successful. We suggest that testing procedures correspond to the goals of instruction. There may be four or five very different types of tests that can be used during a course to evaluate a student's work.

Some of the techniques suggested in Section 4 are also relevant to evaluation, and will not be repeated. Here are some additional suggestions that participants at the conference have tried and found useful:

- open-ended exams (that is, exams with no time pressure, and perhaps open book), where the goal is to test understanding rather than speed. (This can be done with the help of a proctored "test center.")
- some short, time exams, where the goal of the exam is to test understanding *and* speed (e.g., on differentiation).
- exams with problems of graded difficulty: an easy problem or two, some medium problems, some hard ones (rather than having ten problems of comparable difficulty and making the exam a race to the finish).
- formal oral presentations graded as they might be graded in a speech class.
- reading assignments on new material (possibly allowing access to TA's or yourself for answering questions), accompanied by test problems on that material.
- take-home exams, essay questions, oral exams, etc.
- standard questions in "non-standard" formats
- students compile "portfolios" of their best work during the term, illustrating the projects they had completed, problems they had solved, etc.

There are multiple goals for calculus instruction, and it makes sense to design tests with those goals in mind. It is also useful to remind the students explicitly of those goals, and of the purpose of the testing. Developing such a broad variety of tests can be a time-consuming endeavor. We encourage organizations like the MAA to compile "test banks" consisting of a range of evaluation procedures (including sample problems) and discussions of their use.

7 About Priorities and Rewards

Calculus is an instructional "stepchild" in many departments. It is often considered to be a service obligation, with a large captive audience of (somewhat) uninterested students. In departments suffering from enrollment pressures calculus is often the first course to give way to large lectures or to be handed over to TA's. The sentiment, in general, seems to be that a mathematics department's *real* attention should be saved for majors—perhaps beginning with linear algebra courses.

It may be that we are losing some of our potential majors as a result of this attitude and the quality of instruction that results from it.[7] Our service constituencies are unhappy with much of our instruction, and it appears that much of that unhappiness is justified. On the positive side, it is also likely that our own majors, having experienced some "real mathematics" instead of a mechanical plug-in course, would have less trouble in the mathematics courses they take as upper division students. For all of these reasons, it seems appropriate for mathematics departments to devote increased resources to calculus instruction. In short, it is in our best interest to "do it right."

It may be necessary to lobby with college and university administrations in order to obtain adequate resources. Fiscal resources are not enough, however. Earlier we discussed the "bottom line" for students. There is a similar bottom line for faculty. If perfunctory instruction will suffice and if the kind of effort required for truly high quality instruction will go unrewarded, faculty have little incentive to produce such instruction.

We hope that mathematics departments will provide the kinds of rewards that will encourage faculty to teach as well as they can. Moreover, we hope they will adopt the kinds of evaluation procedures that will (a) identify those faculty who are doing well, and (b) help all faculty to improve their teaching. A forthcoming publication from the MAA Committee on the Teaching of Undergraduate Mathematics' Subcommittee on the Evaluation of Teaching will include references for a variety of evaluation procedures.

Statistics indicate that the number of graduating mathematics majors is less than half the number of freshmen who enter college expecting to be mathematics majors. It need not be this way. At one college known for its mathematics teaching, for example, 5% of the entering class declare themselves to be potential mathematics majors—but 10% of the senior class graduates with degrees in mathematics.

8 Issues for Research

This report has been based on information now available to us. This information is adequate to identify some of the problems with current calculus instruction, to suggest the kind of curriculum that would allow students to learn some real mathematics, and to suggest some effective methods of instruction. Yet there is much that we do not know, and that would be most useful as we try to develop better means of instruction. In particular:

We could use more data on current practice. What are current enrollment patterns across the country? What percentage of classes are taught in small sections, in large lectures only, in large lectures with recitations, self-paced, etc.? What kinds of feedback policies are currently used? How often is homework collected and how carefully is it graded? What other kinds of feedback are given? Are placement and diagnostic exams used, and how do the failure rates at institutions with placement and diagnostic programs compare to the failure rates at institutions without them? Do tests such as the SAT Math exams, or the MAA placement exams, serve as good predictors of readiness or success for calculus courses as they are currently taught? Do (or would) they serve as good predictors for the kinds of courses that we would like to see taught? What kinds of testing and evaluation procedures are used? And so on.

We need more information about courses that meet the kinds of goals outlined in this report. What colleges and universities already offer such courses? How does one identify them? How does one document that the courses do indeed meet those goals? How and why do those courses work? What kinds of course structures have been used, what kinds of placement programs, homework and testing policies, evaluation procedures? What kinds of departmental incentives were offered? How did the whole system function?

Finally, we need to know more about what students learn in their mathematics classes. A close look at students' work (by means of interviews, videotapes of students working problems, etc.) is often a disturbing, but valuable source of information. More detailed research on students' mathematics learning would be helpful, both to tell us about current difficulties in instruction and to suggest ways that might help us to improve.

9 Recommendations

Recommendation 1. We believe that calculus instruction can be much improved, and that it can meet the following goals:

- to help students understand a select set of fundamental topics in calculus, and help them develop the ability to apply what they have learned with flexibility and resourcefulness.

- to expose students to a broad range of problems and problem situations, and a broad range of approaches and techniques for dealing with them.
- to help students develop an appreciation of what mathematics is, and how it is used.
- to help students develop precision in both written and oral presentation.
- to help students develop their analytical skills and the ability to reason in extended chains of argument.
- to help students develop the ability to read and use text and other mathematical materials.

We recommend that individual faculty be given freedom to experiment in their instruction, perhaps along the lines suggested in this report. We recommend that mathematics departments allocate adequate resources and establish the appropriate priorities and rewards to encourage high quality instruction in calculus classes.

Recommendation 2. We would like to see the following resources developed and made available.

A. Resources for use with students:

- textbooks and other supplementary materials aimed at presenting the calculus for the appropriate level of conceptual understanding.
- Sets of conceptual problems for all topics in the syllabus, covering broad ranges of difficulty.
- sets of problems that can be appropriately used with the available technologies: number and graph crunchers, symbolic algebra packages, etc.

- a compilation of readings from other disciplines, in which mathematics is used in a significant way.

B. Resources for the faculty's help and guidance:

- Information about successful practices: class structures, feedback procedures, etc.
- Demonstrations of prototypically good instruction: written descriptions and/or videotapes of classroom procedures that have worked, particularly useful problem sets or types of assignments, etc.
- Detailed sets of reliable non-standard procedures for evaluating the students' work, keyed to the goals of instruction.
- Sample descriptions of departmental evaluation procedures, both for identifying and rewarding good instruction, and for helping faculty to do a better job of teaching
- Research results describing current practice. identifying exemplary programs, and pointing the way to improved instruction.

Recommendation 3. We hope that discussions of effective calculus instruction will continue, at meetings of mathematics societies, of faculty and administrators responsible for overseeing their departments' calculus courses, of faculty who teach calculus, and of university administrators.

MAKING IT HAPPEN
Report of the Implementation Workshop

Lida K. Barrett, Chair   Stephen B. Maurer   James R. Stevenson
Andrew Gleason   Peter L. Renz   John Thorpe
    Donald B. Small

1 Introduction

In two days of discussion the 23 conference participants examined ideas, developing some, discarding others, and reached a growing sense of the nature of the mathematics that should be contained in a calculus course, objectives for the course, and changes in style of presentation that should take place in order to reach the objectives. During their deliberations on the third day of the conference/workshop, the seven participants in Workshop III: Implementation articulated needed strategies for implementation of a new calculus course. Because the three workshop groups met simultaneously, the exact content of the new calculus and details of the method of presentation were not known to the implementation panel during its deliberations.

2 Activities Needed for Implementation

Seven sets of activities needed during the implementation process were addressed by the implementation panel, based on the teaching of content to be determined by Workshop I and using the methods and strategies to be described and developed based on the report of Workshop II. The activities discussed below are given essentially as they were addressed by the group as a plan for implementation to be based on existing materials, but each was seen as necessary in an ideal or major revision to be based on new materials. They are grouped under three headings: Development, Field Testing, and Public Relations/Endorsements.

3 Development

Selection of Institution for Participation in Initial Activities. Satisfactory implementation of the new calculus so that it can become the standard will require careful selection of schools to take part in all stages of the project. The participation of prestige or elite schools to validate the project is essential. However, to have the new calculus become the norm will require representation of the breadth of institutions involved in the teaching of calculus—state universities, four-year colleges, community colleges. Participants in Workshop III mentioned Carnegie-Mellon University, Massachusetts Institute of Technology, Harvey Mudd, Grinnell, Williams, Colby, and St. Olaf as institutions who may have already begun to make significant changes in their calculus courses. The MAA *Newsletter*, the American Mathematical Society *Notices*, and the *SIAM News* could be used to raise the questions: Are you currently teaching something different from standard calculus and if so what? Pilot projects might in some cases be parallel to or a continuation of what has already been taking place. In addition to inclusions of innovative schools, schools thought of as conservative in approach should also be included.

A Preliminary Conference of Participants. Once participants are selected, a conference of representatives from participating institutions is seen as essential. The conference would serve the general purpose of dissemination of materials and the development of understanding of these materials. Previously prepared materials should be provided to all participants together with an opportunity to utilize them in practice sessions, followed by activities that require the participants to develop further materials. Instructors will need to learn strategies and techniques to involve students in problem solving. Problems and techniques developed at the conference should be published. In the European style, new problems and techniques should be identified with the name of their authors in order to provide recognition for participants. A network of participants should be established in order to share information and to provide reinforcement and support as the initial projects begin.

Treatment of Material Omitted From The Standard Course In Order To Arrive At The New Course. The goal for the new calculus is an increase in student under-standing of the fundamental ideas and operational meth-ods of calculus sufficient to make unnecessary the exten-sive coverage of topics now presented in a standard calculus text. Evaluation of the success of the new calcu-lus will of necessity be based on how well students can learn and utilize "standard" topics not explicitly covered but needed later. The development of alternative presen-tation techniques and materials (videos, computer as-sisted instruction, small module booklets, omitted sec-tions of the textbook) for topics not covered will be essential. The new calculus course itself should require a student to develop an understanding of topics not covered in class, to learn and be tested on new but related mathe-matics.

4 Field Testing

Implementation Within A Mathematics Depart-ment. At each participating school careful thought must be given to the preparation of the teachers for pilot classes. It is recommended that from the very beginning at schools with graduate programs the group of teachers should include a teaching assistant or teaching assis-tants.

Not only will this establish the creditability of course materials as suitable for general use, demonstrating that they can be taught not only by senior faculty but by new teaching assistants, but the enthusiasm and attitudes of the younger teacher will also contribute to the process. One or more individuals who attended the preliminary workshop should be responsible for a "local workshop." A syllabus for the course, discussions of classroom tech-niques, problems, methods of stimulating discussion, and testing materials should all be reviewed and shared, and sample presentations should be made. Again the ob-jective would be not only to learn about prepared mate-rial but to create material and to practice its use. In a small department it may be desirable or necessary for all members of the department to participate in the pilot program. In a large department it might be the case that parallel sections will be used, some using the new calcu-lus, others continuing past practices. In the latter case, an evaluation can be done of the contrast between the two approaches.

In many institutions the feeling exists that users are aware of the content of the calculus course and the amount of material covered is in response to user de-mand. Especially in these locations, those planning im-plementation must take care to make arrangements for suitable presentation of the omitted materials. (See Item I above.) Hopefully, articles on how to read a mathemat-ics book, how to write a clear explanation, and how to teach the above two topics will be available. The MAA Handbook for teaching assistants, which contains tech-

niques for presentation of mathematics, would be useful.

An essential element in the success of the pilot pro-gram will be a clear understanding of evaluation tech-niques for the program and plans for their implementa-tion during the first year *and* during subsequent years. A full evaluation of the project should be delayed to the sec-ond year.

A serious program, national in scope, for teaching graduate students to teach in the style needed for the suc-cess of the new calculus will in and of itself facilitate the implementation of the new course. If all graduate stu-dents are taught to see the new calculus as the appropri-ate course, if all learn successful methods of presenta-tion, and if this activity can be seen as fun and stimulating, then within a five- to ten-year period mathe-matics instruction will be different. The nationwide build-up of Ph.D. programs caused a greater emphasis to be placed on research and led to a swing away from con-cern with teaching. A new calculus course can provide the impetus for a needed swing back toward more concern for teaching.

Appropriate support for paper grading or other forms of daily work monitoring will cost approximately $15 per student per semester and is seen as an essential resource need. The MAA National Study of Resources for Colle-giate Mathematics will be addressing resource issues broadly. Departments can build on it to address the re-sources needed for a new form of teaching.

Within a mathematics department every effort must be made to keep those in the department not involved in the project aware of the process, aware of its successes and its failures. A fair presentation and an openness will be needed in order for the department as a whole to adopt the project and to support it. It is possible that there will be a rejection of the process by some "old timers" in the department and disarming or neutralizing their opposi-tion can more effectively be done by information, honest statements of what is actually happening—both its strengths and its weaknesses.

Dialogue With Others In The University But Outside The Mathematics Department. It is important from the beginning to emphasize the trade off that is being made—better prepared students who are able to read and work on their own, but a leaner coverage of material. This can be restated by noting that techniques that are not used are promptly forgotten, but concepts that are understood and used in a number of settings will be natu-rally reinforced. Repeated and consistent statements of this theme will be needed to develop a framework in which the pilot program can function and be understood.

It will be politically helpful on a campus to emphasize the change in teaching style and the quality of teaching that will be provided. Certainly the direction of the proj-ect is in keeping with the major study in undergraduate education supported by the National Institute of Education and entitled *Involvement in Learning*. The literature available on learning research and on evaluation of teaching is often unknown to mathematicians and mathematics faculty. It will be important for those involved in the process not only to know this material but to utilize its results in communicating the objective of the new calculus to others outside the department. If there are individuals on the campus who themselves are involved in learning research they can be brought into the project for support.

Opportunities for dialogue with other departments need to be provided on a regular basis. Not only should the mathematics department touch base with other departments as the project begins but an agreed upon method for further contact should be established from the beginning.

5 Public Relations and Endorsements

Public Relations. At every stage of the process there should be a public relation effort. The clear consensus of the appropriate future direction for calculus that developed at the conference should be stated fully in a position paper in order to bring others in the mathematics community to an understanding of and support for the proposed changes. The position paper should be a "puff piece," attempting to communicate the "spirit of New Orleans" to non-attendees. This paper and a news release should be shared with *SIAM News, Focus, Chronicle of Higher Education*, AMS *Notices*, AMATYC, MAA Board of Governors, MAA Section Officers, MAA Section Meetings, the Mathematical/Sciences Education Board, the Conference Board of Mathematical/Sciences. Presentation should be scheduled at meetings of the American Council on Education and the American Association of Higher Education in order to inform administrators of the need and support for change in calculus. Public relations efforts at each stage should be aimed at making the new calculus the property of the entire mathematics community and not just a program of the participants at the New Orleans conference or for those involved in the initial projects. A list of individual and organizational contacts should be made at an early date and progress reports (or perhaps a newsletter) should be issued regularly.

Endorsements. Endorsements of the proposed changes should be sought from key individuals and organizations within the mathematics community, and from individuals and organizations in related disciplines. Obviously the move to obtain endorsements will be heavily dependent upon the success of the public relation effort and of the initial projects. Efforts to obtain endorsements should move gradually and be directly tied to the develop-

ment of firm plans and to initial successes. Endorsements from users of mathematics in other disciplines can provide an important source of encouragement to mathematics faculties and key information to be shared with local faculty in other disciplines as projects develop.

In addition to endorsements, financial support will also be needed. An assumption is made that a primary grant will be provided by a foundation or government agency. However, individual schools could and should build on "local" grants from industry or foundations (e.g. Westinghouse, General Electric, IBM, Amoco Research) or grants available within an institution, from educational agencies, and from regional, state, and local government. Given the expectation of massive change, it is possible that publishers might find a way to assist the overall project financially.

6 Possible Timetables

The implementation panel's ideas did not lead to a formal timetable nor to a consensus on how best to implement the new calculus in an ideal way. A review of the recommendations of all three workshops was seen as necessary before a plan and a timetable could be established. The identification of necessary leadership, the acquisition of adequate funding, as well as time for thoughtful reflection were seen as essential before definite plans could be formulated.

Workshop III's recommendation for a timetable was in an effort to move as swiftly as possible. It was based on the remarkable consensus that took place at the conference and grew out of the enthusiasm of the conference. The panels' initial discussions focused on the development of a plan that would build on the momentum of the conference with the aim of beginning the teaching of the new calculus course in the fall of 1986 in a number of diverse institutions. For this reason a method was sought to avoid the delay that would be necessitated by awaiting the publication of a new text containing not only the content prescribed by Workshop I but also a presentation that would lead to methods and strategies developed by Workshop II. Renz, based on his experience in publishing, estimated that publication of a text of the scope and magnitude for a year or longer course would require at a minimum two years, realistically three. On the other hand, a trial from a set of notes used in a small number of settings might be an insufficient beginning. Building on this hypothesis, the group developed a plan for implementation based on using current texts. A lean coverage of material would be selected from a text. Authors of the best seller texts would be asked for help as they not only know well their own materials but also are experts in writing and organization. This approach could be implemented at a number of schools and could be based on more than one book. Further, comparison between the outcome for sections teaching the new lean calculus and those using the conventional approach would be facilitated.

In answering the question, "Could a truly different content and style be developed from an existing text?" the group identified the following disadvantages. Skipping in a text always creates problems of appropriate sequencing and often leads to inadvertent omission of needed techniques or material. The style of an existing text might make it impossible for the "new" calculus course to seem to be new, not only differing in content and style but also sufficiently different so that the thrust of the new direction would be clear and lead to needed development of new materials and a new text. Additionally, although basing implementation on existing texts has the potential to shorten the time frame, even this strategy might not be possible to implement by fall.

Since the ending of the conference, Ron Douglas has suggested an alternate procedure. He notes that in asking Workshop III to consider the problems of implementation, it was agreed that they had an impossible task since they did not know what they would be implementing. As it turned out most of the effort was concerned with how to implement a partial solution. However, in thinking carefully about the recommendations of Workshop I and II which the group received with considerable enthusiasm, Douglas has come to the conclusion that they are radical and would not fit into a change in the current calculus course. In particular, he does not believe that the recommended course can be taught from existing books. For these reasons, Douglas plans to work toward having pilot programs in place for Fall '87. Presently a prospectus and rough budget have been developed in order to find money from foundations, both private and governmental. The following is a rough timetable.

October 86	Team chosen to prepare textbook and other curricular materials
February 87	Ten colleges and universities chosen to participate in the pilot program
April 87	Working draft of textbook completed
June 87	Conference of leaders of pilot projects and textbook team to make final plans for Fall semester
September 87	Fall semester—ten schools
November 87	Ten more colleges and universities chosen to participate in pilot program
January 88	Conference of all pilot project leaders, old and new, to share experiences and to "fine-tune" the second semester
January 88	Winter semester—ten schools
June 88	Revision of Textbook and other changes based on experiences of the first year
September 88	Fall semester—twenty schools
January 89	Assessment conference and planning conference
March 89	Textbook and curricular materials in final form
Spring 89	Selected workshop presentations to allow additional colleges and universities to introduce the new courses with assistance from the program
September 89	Fall semester—fifty or more schools
January 90	Major presentation at AMS/MAA Annual Meeting
Spring 90	Workshop presentations at all regional MAA Meetings
September 90	Coordination and assistance provided for colleges and universities introducing the reformed calculus for the academic year 90–91

There are many questions left unanswered and the timetable may need changing but "calculus reform for all by 1990" has a nice ring to it.

In conclusion, two aspects of discussion seem to provide a focus for future planning. The first is the unanimity of agreement that the appropriate direction for the future has been found by the conference. The second is contained in the answer given by a participant to the question "Why would anyone try this?". Answer—because it will be more fun to teach the new calculus and this fun and enthusiasm can be conveyed to undergraduates, hopefully leading more students into mathematics as a career and to the rebuilding of mathematics faculties that will be needed in the 1990s.

Papers Presented to
The Conference/Workshop
To Develop Alternative
Curriculum and
Teaching Methods
For Calculus at the College Level
Held at Tulane University
January 2-6, 1986

In Praise of Calculus ●●●

Peter D. Lax, *NYU-Courant Institute of Mathematical Sciences, New York, NY*

Plans for the education of future mathematicians must be based on a shrewd notion about the future development of mathematics. To glimpse the future, we must look at the current state of mathematics as well as the trends in it and in the sciences and technologies contiguous with mathematics.

The single most striking new development in all these fields is the rise, indeed ubiquity, of computers of high speed and large capacity. They make possible numerical and symbolic explorations on an unprecedented scale; much physical experimentation can be replaced by numerical modeling; vast amounts of data can be stored and subtly processed for the extraction of hidden details—as is done, for example, in computerized tomography. This computer revolution has altered the face of applied mathematics. For instance, linear algebra has changed from a moribund to a very active research area because efficient and stable ways of carrying out matrix operations are needed for the numerical solution of partial differential equations.

Tony Ralston is right when he points out that the availability of computing has brought to the fore an impressive array of discrete problems, combinatorial in nature, that are very challenging intellectually and important in a wide range of applications. Some classes of these discrete problems are formalized as topics in computer science, others as topics in mathematics; teaching these latter topics and doing research in them is the task of departments of mathematics.

On the other hand, *it is as wrong as wrong can be to say that calculus has lost in any sense, relative or absolute, its importance in formulating notions and solving problems of mathematics.* The following very brief review of some developments in the last 25 years should set this matter straight:

● The theory of analytic functions of several complex variables and modern differential geometry are two fields that have grown from collections of examples and isolated theorems to edifices deserving the name of theory, with many deep and astonishing results and connections to other branches of mathematics.

● Modern differential topology, dating from Milnor's great discoveries, continues to astonish us with its results and methods. Thom's theory of unfolding of singularities is a thing of beauty and has a great range of applicability. Donaldson's recent work on four-dimensional manifolds is a remarkable combination of Friedman's results and the theory of Yang-Mills fields.

● Modern dynamics succeeded in solving many of the outstanding problems of classical dynamics. The Kolmogorov-Arnold-Moser theorem shows, for example, that the so-called ergodic hypothesis is false in general for differentiable volume-preserving maps. It is now recognized that even low-dimensional systems can exhibit chaotic, pseudo-random behavior. On the other hand, a surprising number of completely integrable systems of physical interest have come to light.

● The modern theory of partial differential equations has a number of impressive achievements to its credit. Using the tools of microlocal analysis—pseudo-differential and Fourier integral operators—a number of problems concerning linear partial differential equations, such as diffraction and scattering, have been solved. The progress in nonlinear theories has been just as great: For example, astonishing things have been discovered about singularities of solutions of variational problems, (in particular, about minimal surfaces) in more than 7 variables.

Some important partial differential equations have been recognized as infinite-dimensional, completely integrable Hamiltonian systems, and their solutions display unusually stable structures, called solitons.

• The theory of probability has made great strides; among its modern tools is integration in function space, introduced by Wiener and used by Feynman, Kac, and many others in recent times in a variety of contexts.

• The branches of applied mathematics which might be called classical have not lagged behind. In fluid dynamics, we are beginning to understand much better the generation and propagation of shock waves in compressible media. For incompressible media, we have learned to estimate the size of possible singularities, and also the size (in the sense of Hausdorff dimension) of the so-called strange attractor sets. It is likely that these developments, combined with notions of modern dynamics, will bring us one step closer to understanding the mystery of turbulence.

• Computational fluid dynamics is almost entirely a modern creation. Its tools are finite differences, finite element approximations, spectral and pseudospectral methods, Monte-Carlo techniques, and off-beat ways for treating vorticity and discontinuities. The consumers of computational fluid dynamics are the aerospace technologies, meteorologists, oceanographers, astrophysicists and others.

• Two other active applied fields are quantum field theory and statistical mechanics. Lately, these two appear to merge inasmuch as methods developed in one of them seem applicable to the other. In this connection, I mention a statistical mechanics approach to discrete problems: the method of simulated annealing developed by Kirkpatrick, Gelatt, and Vecchi (*Science*, May 13, 1983), and applied by them very successfully to the Travelling Salesman Problem.

• A neoclassical applied branch is mathematical biology and physiology; the father of the field is Helmholtz, but large scale computing is an important ingredient of its recent successes.

All of these recent developments are in calculus-based branches of mathematics. To underemphasize calculus during the formative years of future mathematicians would prevent many from acquiring the kind of facility with calculus that is needed to work in analysis—pure, applied or mixed. *A calculus-deficient education would shunt students into a small corner of mathematics, instead of opening up its whole panorama.* Fortunately, it is very unlikely that such a one-sided curriculum would be adopted in a department that has a balanced view of mathematics.

I am not saying that finite mathematics should not be taught; it should—to mathematicians and computer scientists. It seems reasonable to try to introduce some of its topics into the last year or so of the high school curriculum.

As to calculus: mathematicians need not less, but more of it. The real crisis is that at present it is badly taught; the syllabus has remained stationary, and modern points of view, especially those having to do with the role of applications and computing, are poorly represented.

Reprinted with permission from <u>The College Mathematics Journal</u>, Volume 15, Number 5, November 1984. Copyright 1984 by the Mathematical Association of America, 1529 Eighteenth Street, N.W., Washington, D. C. 20036.

The Importance of Calculus in Core Mathematics*

Ronald G. Douglas

Ronald G. Douglas is a professor, Dept. of Mathematics, SUNY at Stony Brook, Stony Brook, NY 11794.

Although calculus has formed the core of the undergraduate mathematics curriculum for most of this century, there has been much debate recently concerning this role [3]. The case is being made that discrete mathematics is now of paramount importance and should therefore form the core. Discrete mathematics, the mathematics dealing with discrete finite systems, includes such topics as the study of permutations, Boolean algebras, matrices, semigroups, and graphs; it considers both their structure and their manipulation. Although the study of such topics goes back a century or more, recent interest has been sparked by the growing importance of computers in our society.

Calculus has been a central factor in our expanding knowledge of the universe. It is the key to understanding systems that change in the social, the biological, or the physical sciences. Students need calculus to understand the nature of scientific laws and their application in the modern world. Rapid large-scale computing has increased, not lessened, the role of calculus in solving many of the outstanding problems of science and technology. The issue should not be whether to replace calculus in the core curriculum, but rather

how to transfuse and invigorate the pallid, passive calculus which is presently taught in so many American colleges and universities. To speak of a "calculus crisis" would not be overly dramatic.

Calculus has stood the test of time and has been taught to science and engineering students since formal study in these areas began. It is the foundation and wellspring for most modern mathematics. Moreover, the development and application of continuous mathematics continues unabated. Indeed, its applicability is enhanced and strengthened by the advent of high-speed, large-capacity computers. In a recent note, "In Praise of Calculus" [1], Peter Lax summarized a few recent striking developments in the mathematical understanding of dynamical systems in physics, of scattering and diffraction in wave phenomena, and of the generation and propagation of shock waves in fluid dynamics.

These are just a few recent developments in calculus-based branches of mathematics. It is clear that mathematicians, both pure and applied, need to learn calculus early in their careers. Similarly, computer scientists will need to learn calculus if they are to understand many of the most exciting applications of large-scale computing.

The development of calculus coincided with the revolution in the physical sciences in the seventeenth and eighteenth centuries. It is no coincidence that Newton is honored both as a codis-

*This is based on my presentation on the panel entitled "Discrete Mathematics as a Rival to Calculus in the Core of Undergraduate Mathematics" during the annual meeting of the American Association for the Advancement of Science held in New York, May 25, 1984.

coverer of the calculus and as the founder of Newtonian mechanics. The understanding of physical laws is inextricably interwoven with calculus. The difficulties inherent in the well-known physics courses without calculus make this clear.

Almost all of science is concerned with the study of systems that change. The analysis of such systems is the very heart of the differential calculus, and indeed that analysis cannot progress very far without it. The description of such systems usually takes the form of an ordinary or a partial differential equation which the system satisfies. The numerical solution of such equations is one of the principal tasks of large-scale computing. Even the analysis of such equations using finite differences is next to meaningless without an understanding of the calculus. Thus, all science and engineering students need calculus very early in their studies.

But calculus is useful for a much larger group of students. Two decades ago C.P. Snow described the "two cultures" phenomenon, that is, that modern educated society can be divided into those people with an understanding of science and those without. The consequences of this division continue today, and Snow's cultural grouping can be sharpened into those people who understand the role of mathematics in explaining and studying our world and those who don't. Calculus is an excellent laboratory in which all students can begin to learn this important role of mathematics.

The case for calculus remaining in the core of undergraduate mathematics is overwhelming for a wide variety of students. If that is so, then why is it being questioned, and what is the fuss all about? In part, it is because calculus is widely perceived as being difficult; in part, it is because calculus has not been rethought since the advent of computers; and, in part, it is because there is a large body of useful discrete mathematics that doesn't fit conveniently into the present curriculum. And finally, it is because calculus is not being taught very well in most American colleges and universities.

Calculus is difficult because calculus is difficult. It is a coherent theory that builds on all of high school mathematics and then builds on itself. That is, one must thoroughly understand what has come before in order to go on. The calculus student is attempting to understand the culmination of humankind's grappling with the notions of limit and of the continuum. This stretches and enlarges the students' physical and geometrical intuition. It demands that they master techniques that are two and three centuries old, but still useful. These ideas are worth struggling over, and there is a pay-off.

The structure of today's calculus course has changed little in the last several decades. Although various reforms have been proposed from time to time, a quick review of the most popular calculus texts will show that little change has occurred. Many topics are included because an individual department wants it, and most attempts at changing the standard calculus courses have failed. Certainly, the time is overdue for a serious rethinking of the calculus we teach. I hope that will be one of the outcomes of the debate on calculus versus discrete mathematics.

There is widespread interest in many topics from discrete mathematics; they are useful in computer science and many of its supporting activities. Moreover, there are many exciting and interesting applications of discrete mathematics, and many areas are flourishing. Obviously, some students need to learn some or all of these topics, but not necessarily all students, and it should not be at the expense of calculus.

Moreover, the argument for any of these topics being in the core is not convincing, especially since it is difficult at this time to decide what will stand the test of time and what won't. In particular, no coherent discrete mathematics course exists. Rather, a series of topics continues to evolve, and the precise list depends on the person proposing it. Such courses are important in the undergraduate curriculum, but play a limited role in the *core* undergraduate mathematics curriculum.

Finally, the principal problem with calculus today is the way it is taught in most American colleges and universities. Calculus is usually taught in large lectures, with very limited interaction between students and teachers. Moreover, homework is only sometimes assigned and almost never graded. This is not because instructors today are lazy or less interested in students' learning calculus, but because of numbers. An instructor teaching 200–300 students cannot collect and read homework; there is barely time to give and grade two or three exams a semester. Moreover, even recitation sections have grown to the point where teaching assistants can do little to ameliorate the situation.

The demoralizing effect this has on both faculty and students cannot be overstated. Calculus cannot be learned passively. As the subject builds, the student must continually master ideas and techniques in order to profitably continue [2]. Therefore, it is not surprising that the goals I stated earlier for calculus are hardly ever achieved. This is the real crisis in core mathematics, and our energies should be directed to solving it. ∎

References
1. Lax, Peter. "In Praise of Calculus." *College Mathematics Journal* 15(5):378–80, 1984.
2. Macroff, Gene I. "Class Size is Key to Campus Success." *New York Times*, Feb. 26, 1985; pp. 17–18.
3. Ralston, A., and G.S. Young, Eds. *The Future of College Mathematics*. New York: Springer-Verlag, 1983.

PROPOSAL TO HOLD A CONFERENCE/ WORKSHOP TO DEVELOP ALTERNATE CURRICULA AND TEACHING METHODS FOR CALCULUS AT THE COLLEGE LEVEL

Submitted by:

Professor Ronald G. Douglas
Dept. of Mathematics
State University of New York at Stony Brook
Stony Brook, NY .11794-3651

Telephone #: 516-246-6522

A Proposal to Hold a Conference/Workshop to Develop Alternate Curricula

and Teaching Methods for Calculus at the College Level

1. Introduction

The importance of calculus in the college curriculum has long been recognized. Calculus has been taught to students in science and engineering since formal study in these areas began. More recently, students in the biological and social sciences have been required to take a semester or even a year of calculus. The percentage of undergraduate students enrolled in calculus has grown steadily during the past two decades and now at most American colleges and universities, considerably more than half the students take calculus sometime during their studies.

Since calculus is central to the understanding of systems that change, whatever the field, it has played a central role in expanding our knowledge of the universe. Further, although the mathematical sciences have grown and burgeoned in all directions, calculus has retained its centrality in the core undergraduate mathematics curriculum. Finally, performance in calculus courses plays an important role in determining those students who will go on to pursue careers in science and engineering and even in professions such as medicine. In fact, poor performance in calculus is one of the larger obstacles to obtaining

---

1 Attached are prepublication copies of two articles which further elaborate on the importance of calculus.

a better re-presentation of minorities and women in these
careers. For all these reasons, the successful study of calculus
is of critical importance to our society.

Despite this, the teaching of calculus is in a state of
disarray and near crisis at most American colleges and universities.
Evidence of this is provided by a failure rate of nearly half at many
colleges and universities. Moreover, when the current level of performance of even the successful students is compared with that of students
two decades ago, one really understands the cause for concern. Morale
is low among both faculty and students and nearly all faculty now view
the teaching of calculus as an unpleasant but necessary chore. While
there are exceptions, they are few and far between. At the recent
(well-attended) AMS/MAA Panel entitled "Calculus Instruction, Crucial
But Ailing" held in Anaheim on January 11, 1985, no one took issue with
the statement that "calculus instruction is ailing" and each speaker,
both from the panel and from the audience, had a personal "horror story"
to tell.

The problems with calculus instruction are serious and many.
(1) The calculus curriculum has not been seriously rethought for over
two decades and certainly, not since the ubiquity of computers
available for both numerical and symbolic computation. The present
calculus curriculum contains too many topics so that important ideas
and concepts are often not understood or are overlooked even by the good
students.

(2) The present mix of calculus students is drastically different
from that when the current curriculum and teaching methods were devised.

(3) The level of expectation in most calculus courses was lowered to
accommodate the flood of poorly prepared and poorly motivated students
which followed the "sixties." Many students are capable of much more
and both students and faculty realize this. Moreover, this low expectation
has allowed performance in high school mathematics to deteriorate further.

(4) Finally, calculus enrollments have grown dramatically during the
past decade as have all undergraduate mathematics enrollments, while
resourses in many instances have shrunk. This has resulted in larger
classes with less and less interaction between students and instructor. Although this is widely understood for offerings in computer
science, it is actually true for all the mathematical sciences.

As we mentioned earlier, the existence of serious problems in
calculus is well-understood within the mathematical sciences. Hence
attempts to improve calculus instruction have been made by mathematics departments at many colleges and universities as well as by
several national organizations in the mathematical sciences. For the
most part, these efforts have had little impact on the problems.

The difficulties inherent in one department making significant
changes in calculus instruction are obvious. A course as large as
calculus needs a textbook and the only textbooks available present
the standard curriculum. Moreover, since departments in other
disciplines want the topics "essential" for them to remain unchanged,
this leaves little latitude for reform. Finally, the faculty in most
mathematics departments have such diverse opinions on the calculus,
it is difficult to reach any consensus for change. The latter is what

makes it unlikely that any recommendation for far-reaching change
will originate from within any organization representing the mathematical sciences. For example, although the recent MAA Curricular Study
in the Mathematical Sciences had a panel devoted to calculus, the recom
mendations which it made were relatively bland and almost uniformly
ignored.

I believe a different approach might yield significant results.
I propose that the Sloan Foundation support a small intensive
conference/workshop in January, 1986 at Tulane University in New
Orleans. Representatives from all the mathematical sciences would be
invited with emphasis on choosing people with a long standing interest
and expertise in calculus. Spokesmen from those disciplines which
depend on calculus would also be sought. Finally, people with actual
experience in the teaching of calculus would be invited looking especially for representatives of currently successful innovative
programs. The participants would be asked to formulate alternate
curricula and methods for teaching calculus. Further, the participants
would be asked to address the very real problems, both political and
economical, in having the proposed alternate curricula tried and adopted
at American colleges and universities.

This would be a formidable task but one that is important and
infinitely worthwhile. The widespead agreement that there are serious
problems in calculus instruction which must be addressed, has, I
believe, provided a real opportunity for the success of this approach.
And the support of the Sloan Foundation would give the recommendations
of the proposed conference/workshop a visibility and a credibility which
would be difficult to achieve in any other way.

2. The Proposal

a) We propose that a four day conference be held in January, 1986 on the campus of Tulane University in New Orleans (in conjunction with the annual joint meeting of the American Mathematical Society and the Mathematical Association of America to minimize travel expenses).

b) Format of the conference:

i) A kick off dinner with a provocative keynote address held held the night before the conference begins, followed by informal discussions and an opportunity for the participants to get acquainted.

ii) Two days of prepared papers (see below for topics) with substantial time for the discussion of each paper.

iii) One day of group workshops to develop position papers on various topics (see below).

iii) A final day to discuss the position papers and to plan future steps.

c) Number of attendees: up to twenty-five with most involved as either presenters of papers or workshop leaders.

d) Topics to be covered:

The following list is tentative and likely to be changed as participants are chosen and add their ideas and points of view. Papers would be prepared and distributed to the participants before the conference.

i)   What are the essential aspects of calculus that a student should learn in the standard calculus curriculum?   Should there be more than one calculus track at a small college? at a larger college or university? What should the different emphases be? What topics in the current curriculum should be scaled down or even omitted?

ii) What is the appropriate level of rigor for the standard calculus sequence(s)? Should the emphasis be on rigor, in developing intuition, or in providing a conceptual understanding? How important is the role of calculus in developing mathematical maturity?

iii) How can the student best learn that calculus is the study of systems that change? Can one understand the role of calculus in scientific thought without including applications in calculus courses? Is it important that students understand this role of calculus?

iv) Computer software is now available which will solve the routine problems of calculus such as the differentiating and integrating of elementary functions in closed form. Should as much class time be devoted to learning to do such calculations as has been standard in the past? Should more stress be placed on understanding the applications of these techniques?

v) Calculus can be viewed as the study of linear (and quadratic) approximation of "nice functions." This fits in well with numerical analysis and can be nicely illustrated with the aid of the computer. Should this be done in calculus? For all students?

vi) Calculus is a course which builds on itself, needs constant reinforcement of learning, and in which rote memorization has little role. Can calculus be taught effectively in large-lecture sections? Is the collection and reading of homework necessary? What is the best use of teaching assistants, either graduate and undergraduate? Is personal contact with the lecturer important?

vii) Are there enough contact hours to do justice to the present curriculum or to alternate curricula? In high school, calculus classes are small and there are five to ten class hours a week. Would many students benefit if calculus were taught under similar conditions in college?

viii) Everyone is dissatisfied with the current crop of calculus textbooks. Yet, if an author writes and manages to get published a textbook which is a little different, most colleges and universities will refuse to use it. How can we break out of this cycle?

ix) Many of the best students have taken calculus in high school. What should be done with these students? Should there be an "honors calculus"? For whom? And what should the emphasis be?

x) How can one identify currently successful innovative calculus programs? How can such programs be transferred and adopted at other colleges and universities?

xi) What would be the major problems in implementing new calculus curricula or a new approach to calculus instruction at a college or university? How can they be overcome?

xii) Any change in calculus is likely to produce positive effects, both in performance and morale in the first few years. Can the long run effectiveness of a new program be predicted in the short term?

e) Third-day Workshop Topics

The papers presented on the first two days will reflect a variety of viewpoints and will undoubtedly spark some controversy. It will be the task of the third-day workshops to develop concrete proposals for alternate curricula and teaching methods. Although more than one proposal is likely, all proposals should represent the recommendation of the conference. The workshops will attempt to synthesize the ideas in the various papers into proposals.

Workshop 1: Develop one or more detailed college calculus curricula.

A proposal should include a detailed list of topics to be covered in one or in a sequence of courses, the philosophy of the course, the intended audience, the goals, and the intended approach.

Workshop 2: Develop recommendations on teaching methods for calculus.

The workshop should prepare a critique of existing teaching methods stating both the strengths and weaknesses of a given method for various groups of students. New and innovative methods should be described in enough detail to allow them to be tried.

Economics and other factors should be addressed but the final recommendations should not be bound by these factors.

Workshop 3: Develop a position paper on how to proceed after the conference.

This is especially important since many previous curricular reforms for calculus have had little effect because of inadequate follow-up. The workshop should address the question of how the recommendations of the other workshops can be implemented. How can colleges and universities be induced to try the proposals? How can the results of the trials along with the other recommendations of the conference be made available to the mathematical sciences community and other interested parties.

f) Final Day

i) Morning - reports by the three workshop leaders and discussion of the reports.

ii) Afternoon - general session to discuss any remaining issues, loose ends, and to summarize the conference/workshop.

OPENING REMARKS AT THE CONFERENCE/WORKSHOP ON CALCULUS INSTRUCTION

R. G. Douglas

Let me begin by welcoming you all here. The problems of improving or even changing an enterprize as large as that of calculus instruction are daunting but I'm excited and grateful that you have agreed to help me try.

I first encountered the ideas of calculus in my reading of popular physics literature in about the ninth or tenth grade. I remember well my frustration at my inability to share my discovery with my classmates although try I did. I am reminded of that experience every time I teach calculus. Even so I would not have imagined myself in the role of calculus reformer even a couple of years ago. Let me say a few words on how I happen to be here.

Most of you are familar with the current debate on the roles of "continuous" and "discrete" mathematics in the undergraduate curriculum. I joined that debate on the side of calculus a couple of years ago. Early on in my preparation I decided I could not defend calculus as it is presently taught but rather I had to speak in terms of what it could be. In thinking about this I began to realize just how far from such an ideal we have come in most colleges and universities. Further, as a department chairman I was already only too familar with the "failure rates" and the other "nuts and bolts" issues affecting calculus. My mind began to search for some way to change the situation. After meeting and talking with Steve Maurer and Cathleen Morawetz, one an officer and the other a trustee of the Sloan Foundation, I came up with the idea for this conference. Because I felt the problem to be an extremely important one and that this approach had a chance for success, I organized this conference and managed to persuade all of you to join me.

I am not going to try to defend calculus here. Some of my thoughts on that issue appear in the article appended here. And the article of Peter Lax, appended to his position paper, also speaks to this issue. Rather we shall accept at this conference that calculus is important! Our task is, rather, to rethink what and how calculus is taught at American colleges and universities

and then to reaffirm that which is good and recommend changes in
what is not. When more than one possibility seems to us to be
valid, then we should say so. In preparing for this Conference, I
reread the recommendations of the MAA Committee on the Undergraduate Program in Mathematics for calculus. As clearly indicated
in the report itself, the subcommittee on calculus made conservative recommendations but recommendations which I believe are
sound. Unfortunately, I don't believe the report has had much
effect.

In organizing this conference I have held back somewhat in
presenting my own thoughts and ideas. I haven't invited you here
merely to ratify a proposal already formulated by me to save
calculus. Many views and opinions are represented here and I
expect we will hear at least two sides to every issue. Still I
want to take this opportunity to share some of my thoughts and
opinions. Although I can't document their validity since most
are based on only anedoctal evidence, they seem to fit the known
data well and there has been substantial agreement by those with
whom I have discussed them.

What are the problems with calculus instruction? Some of
you have suggested in your position papers that no problem exists
and indeed in discussing calculus over the past few years with
colleagues across the country at different kinds of institutions
I found that the situation of calculus varies greatly from college to college. Still at most places, teaching calculus is
viewed at best as an unwelcome chore by both young and old faculty. The feeling is that one is not teaching "mathematics" however one defines it. Indeed, this is borne out by the fact that
few, if any, undergraduate students are inspired to major in
mathematics as a result of their calculus course. Actually the
contrary is often true--one sees students who decide not to major
in mathematics because of their calculus course. When I was an
undergraduate in the late fifties, the hot areas were engineering
and the physical sciences, but in most calculus classes one or
two students became sufficiently inspired by the beauty and the
utility of mathematics to become mathematics majors. We lament
the legions going into computer science today but would we have
gone into mathematics on the basis of the present calculus
course.

There are many causes for the problems in calculus instruction. Let me discuss a couple. First a larger and different mix

of generally less well-prepared students take calculus today.
This is a consequence of major societal forces and changes in our
country and we could devote several conferences to discussing
this but that is not my intenion. We get the students we get and
we should formulate calculus instruction for them. We should not
try to design it for either superstudents or superteachers but we
must work with what we have. We must design calculus for the
needs of the students who take calculus. We should not, however,
water it down. We must not apologize because calculus is difficult but we should make clear our commitment to the integrity of
the subject. I believe we should concentrate on offering a
conceptually oriented course in which many secondary topics have
been eliminated from both semesters. I believe students should
learn that the fundamental notion of the differential calculus is
how effective linear and quadratic approximation is for studying
nice functions and how this can be used to study systems that
change. Students should see in calculus how mathematics is used
to understand the real world. We should describe no more than
two or three syllabuses for each of the first two semesters of
calculus. We should do this in sufficient detail as to topics,
appplications, definitions, proofs, etc. to make our intentions
clear. We want to discuss the emphasis of the course--the kind
of problems the student is expected to be able to do, whether a
conceptual or a rigorous understanding is expected, etc. Many
participants make clear in their posiĝtion papers that they
believe that how we teach calculus is at least as importatnt as
what we teach.

Second, we have tried to teach calculus as though it could
be learned passsively with little contact between student and
instructor. This is largely because we have been unwilling or
unable to communicate with administrators, politicians, and the
general public on the special needs of calculus instruction. The
subject builds, it uses all of high school mathematics, and it
requires constant interaction with questions asked and answered,
and homework collected, read, and returned. Calculus can not be
taught with large lectures and large recitations only. Students
must feel that there is someone interested personaly in their
progress. A student in calculus who waits for the midterm to
study is almost certainly doomed to failure. If a student does
poorly on one topic in calculus, he will not have another chance
on the next, a situation which is common in introductory courses
in most other subjects. Deans, provosts, and presidents must
understand this, as well as governors, legislators, and the

general public. It is impossible to do calculus well without adequate resources, and a large part of the calculus problem is that financial stringencies during the past decade or more have forced many universitites and colleges to try.

To summarize we must agree on at most two or three sylla-buses for the first two semesters of calculus. This must include coming to grips with the role of the computer in calculus al-though not all options need to treat the computer the same. We should recommend possible styles and approaches. Moreover, we should be honest in discussing the resources needed to teach calculus effectively. Perhaps we will need to test our recommen-dations with experiments. Finally, we must address the problems of making certain that our recommendations are discussed, modi-fied if necessary, and then broadly adopted. This requires confronting the very real economic and political issues surrou-nding calculus instruction. Lastly, we have to figure out how to put fun back into teaching calculus--how to give faculty an enthusiam for teaching calculus which at present is largely absent. All of this is indeed a tall order and this Conference, no matter how successful can only be the beginning.

The Role of Calculus in a University Curriculum: A Case Study

Lida K. Barrett and Kay Van Mol

Introduction

A systematic analysis of the mathematics references in the undergraduate catalog of a college or university coupled with an appraisal of the roles mathematics plays in baccalaureate degree programs brings into focus the role of calculus in undergraduate education. The contrast between the beauty and power of calculus and the mastery level required of undergraduates leads to recommendations for a modified and enhanced presentation of calculus. Such an analysis of the calculus component of undergraduate education at Northern Illinois University follows.

Mathematics as a General Education Requirement

Northern Illinois University is a comprehensive university with colleges of Liberal Arts and Sciences, Business, Education, Visual and Performing Arts, Professional Studies (predominantly health-related--nursing, physical therapy, community health, communicative disorders, medical technology--but also home economics and library science), and a new College of Engineering and Engineering Technology which is being developed based on an engineering technology program. All undergraduate students are required to complete a common general education program[1] of 37 hours, including 12 semester hours of university requirements (6 hours of English, 3 of communication studies, and 3 of mathematics). The mathematics courses that satisfy this requirement are College Algebra, Introduction to Mathematics (for the development of mathematical skills useful in daily life), Trigonometry and Elementary Functions, Foundations of Elementary School Mathematics, Finite Mathematics, and the first semester of a three-semester sequence in calculus. Additionally, as a part of general education[1] a student must earn 7 hours in courses chosen from a specified list of courses in biological sciences, chemistry, computer science, geography, geology, mathematical sciences, physics, and statistics (taught by the Department of Mathematical Sciences). The mathematics courses that meet this requirement are a basic statistics course (not calculus based), entitled Excursions into Mathematics (for the development of a student's understanding and appreciation of selective mathematical concepts) and the first semester of calculus (which cannot be used to meet both the university requirement and science requirement). Mathematics majors cannot use mathematics courses to meet the science requirement.

Mathematics Used as a Prerequisite or Corequisite

Mathematics courses appear in the curriculum structure most often as prerequisites or corequisites for courses in other departments. At Northern Illinois University, 13 different mathematics courses are

21

used as prerequisites for 67 courses. Six of these courses--College Algebra, Introduction to Mathematics, Trigonometry and Elementary Functions, Foundations in Elementary School Mathematics, Finite Mathematics, and Basic Statistics--do not require an understanding of calculus. These courses are used as prerequisites for non-calculus-based science courses such as general chemistry, general physics, and astronomy. The basic statistics course, or a calculus-based statistics course, is a prerequisite for two sociology courses and two health science courses. At Northern, as at many universities, statistics courses are offered in the College of Business and in some departments in other colleges (e.g., in the College of Education). There are 19 courses for which one of the five pre-calculus mathematics courses is used as a prerequisite. Finite Mathematics is a prerequisite or corequisite for several courses related to data processing, business statistics, and mathematical modeling in the social sciences. Mathematics for Elementary Teachers is required for courses in elementary education and special education.

A one-semester calculus course for business and social science serves as a prerequisite for only two courses, Computer Programming in FORTRAN and Introduction to Mathematical Methods for Economics. For the economics course, two semesters of the three-semester calculus course is an alternate choice as a prerequisite. The business and social science calculus, however, is an elective sometimes taken prior to entry to the junior/senior programs in the College of Business, and as an optional course that may be used toward satisfying the Bachelor of Science degree requirements in the College of Liberal Arts. (These degree requirements are described below.) The three semesters of calculus serve as prerequisites as follows: Calculus I for 16 courses; Calculus II, 14 courses; Calculus III, 4 courses. Departments requiring Calculus I and II include biological sciences, chemistry, computer sciences, economics, engineering technology, geography, geology, and physics. (The use of the calculus sequence will increase as the engineering program develops).

In the Department of Mathematical Sciences, Foundations of Applied Mathematics, Differential Equations, the calculus-based statistics course, and Introduction to Probability and Statistics all require a calculus background. The statistics courses are required by the social sciences--economics, geography, psychology, sociology, political science--and by geology. Differential equations and applied mathematics courses are required by chemistry, physics, and technology. When a mathematics course is stated as a prerequisite for a given course the implication is that the specific content of that mathematics course is needed for satisfactory completion of the course.

The College of Liberal Arts and Sciences requires all students earning the Bachelor of Science degree to demonstrate competence in laboratory science and/or mathematics/computational skills equivalent to that obtained through two years of regular college instruction (11-14 semester hours). There are four alternatives for completing this requirement: the first includes Finite Mathematics, Calculus for Business and Social Science, Elementary Statistics, and a computer science programming course; the second, Calculus I and II and a computer science programming course or Introduction to Probability and Statistics; the third, Finite Mathematics, Calculus for Business and Social Science, and a two-semester laboratory science sequence; the fourth, Calculus I and a two-semester laboratory science sequence. Alternatives one and two are required for a Bachelor of Science in major programs in communication studies, for certain programs in computer science, and for majors in economics, geography, history, political science, psychology, and sociology, as well as for a comprehensive major in the social science. Alternatives three and four are required by biological sciences, chemistry, certain programs in computer science, geology, meteorology (a geography major), and physics. Any of the four alternatives are acceptable for anthropology, the general major in communication studies, and journalism.

The College of Business offers a junior/senior program with ten freshman/ sophomore courses required of all students prior to admission in the junior year. These courses include Finite Mathematics and Business Statistics (taught in the College of Business).

The College of Education offers only three undergraduate degrees: Elementary Education, which requires two mathematics courses especially designed for this program, Special Education, which requires the same two courses, and Physical Education with no requirement in mathematics beyond the university's general education requirements. The university's general education requirement in mathematics satisfies the mathematics portion of the State certification requirement of ten semester hours of mathematics and science.

In the College of Professional Studies the professional programs have science requirements in chemistry, biological sciences, and physics. A student may elect the non-calculus-based courses and in that case needs only College Algebra and Trigonometry and Elementary Functions. Certain of these programs also require Basic Statistics. Courses in home economics (now called human and family resources) also require only pre-calculus courses with certain programs requiring Basic Statistics.

The College of Visual and Performing Arts, with Departments of Art, Music and Theatre, requires no mathematics (beyond the general education requirements) in any of its programs. In the new College of Engineering and Engineering Technology the standard three semesters of calculus as well as differential equations will be required. The technology programs in the college will require only one semester of calculus.

The Three Roles of Mathematics

Mathematics courses serve <u>three</u> roles in the curriculum. First, mathematics is required for all students seeking a baccalaureate degree. This acknowledges the pervasive role of mathematics in our society by requiring all students to develop an understanding of the concepts and modes of thought of mathematics. Second, mathematics courses are used as tool courses for other disciplines. The set of mathematics courses required in social science programs--introduction to calculus, computer programming, and statistics--acknowledges the need of students in these programs to have certain mathematical tools for work in their disciplines. In the science programs, calculus, differential equations, and applied mathematics are tools needed for success in those majors. Third, mathematics courses function most often in the curriculum as prerequisites for given courses. In these cases mathematics is often a language used to state concepts of the discipline, to develop new results, and to solve problems. A student without the particular mathematical skills required for a given course presumably cannot understand the material of the course.

The Contribution of Calculus to the Curriculum

There are portions of curriculum for which no calculus is required. At Northern Illinois University these include programs in the College of Visual and Performing Arts, the College of Education, the Bachelor of Arts programs in the humanities, and most of the degree programs in the College of Business where the stated requirement is finite mathematics and non-calculus-based statistics. For these students the knowledge, understanding, and appreciation of mathematics is limited to basic mathematical concepts not including the concepts of calculus.

The Bachelor of Science degree programs in the social sciences and the sciences include in all cases a calculus requirement. In the social sciences a one-semester course of an elementary nature in calculus presents the ideas and techniques of polynomial calculus with an explanation of differentiation and integration. The theorems that establish results and the theoretical framework are sketchily presented. The science programs require the three full semesters of calculus, and in most cases a student is introduced to the foundational theorems of calculus.

The use of mathematics courses as prerequisites and corequisites for other courses in specific degree programs gives the best evidence of the importance of the concepts of mathematics and calculus to a given discipline. Science courses require mathematics in order to present results at an advanced level in a systematic way. The ideas of calculus are used in a substantial manner. However, the science courses required in programs in the health sciences--nursing, physical therapy, medical technology--are basic science courses and carry no calculus prerequisite.

The knowledge of mathematics, mathematical ideas, thought patterns, and techniques is seen as valuable to the social science student and a one-semester calculus course is required in each program. Calculus, then, is required in the curriculum of these degree programs, but the techniques of calculus are not seen as sufficiently needed in course work to require a calculus prerequisite except in a single elective course in research methods. The calculus course can be, and often is, postponed until the senior year.

Conclusions

To a mathematician this review of the overall picture of the mathematics requirements presents no real surprises. The thorough incorporation of mathematics within disciplines outside of science has not taken place. There is much talk of the growing quantification of society, and science and technology increasingly affect our lives. For example, statistical techniques are often used to arrive at projections (e.g., in elections) and to present information. In spite of these facts, the ideas of calculus--rates of change and summation, their interrelationship, and the computation and assessments that can only be done with calculus--are beyond most educated individuals' reach.

Calculus as it is known and appreciated by mathematicians plays a minor role in the overall curriculum. Scientists and engineers learn the techniques of calculus and through their work learn to appreciate some of the power of the ideas of calculus and the way these ideas interplay with those of their given disciplines. However, most other students take a calculus course that teaches predominantly computational skills. They only learn calculus techniques that can solve problems--rates of change, areas, and certain kinds of summations.

Students best learn to understand calculus and retain their understanding if future coursework requires the understanding of the ideas of calculus in order to use those calculus techniques. This element is missing for most students studying outside the sciences.

Questions

What does calculus have to offer to the overall curriculum? What could be the nature of a calculus course designed for the social sciences or the humanities? What might be the impact of a calculus course that would enable educated citizens to more fully understand the nature of calculus, the power of its results and ideas, and the effect that the mathematics based on calculus has had on science and technology? Would a wider appreciation of calculus expand an educated citizen's appreciation of our technological society?

Recommendations

The average university student today, given the level of high school preparation, is not prepared to take a rigorous calculus course without a sequence of preparatory courses. The presence within the curriculum of a calculus course that presents problems that can be solved only by calculus and that presents the concepts of rate of change and of summation, and the relationship of rate of change to summation would provide a view of quantitative results and of possibilities for problem solving that is not present in today's curriculum. A calculus course that presents the ideas of calculus in a way that is accessible to more of today's students, one which will lead to an understanding of these ideas as more than formulas that lead to numerical results, could expand a student's understanding of science and the scientific results possible only after the ideas of calculus are known. An appropriate treatment of calculus, a course required of all undergraduates, might lead to a better understanding of aspects of the scientific revolution and the widespread impact of science and technology in the contemporary world.

Most current calculus courses attempt to accomplish these objectives. However, until the educational process requires a further use of calculus beyond an introductory course, students' command of the ideas of calculus cannot be assessed or reinforced. Not only must we write better textbooks for calculus courses, but we must also see that writers in other disciplines present results in such a way as to display directly the contribution of mathematics, and in particular calculus, rather than in a manner that avoids these ideas and thereby obscures their contribution. We must train scholars in other disciplines who understand and appreciate mathematics. As "writing across the curriculum" has come to be seen as necessary to develop students' power of critical thinking, "mathematics across the curriculum" is essential if we are to produce educated citizens not intimidated by science and technology.

[1] A new general education program recently adopted will require 41 hours. The mathematics requirement will be based on two years of high school mathematics and college algebra will no longer meet the requirement.

CALCULUS AT UNIVERSITY HIGH SCHOOL

Robert B. Davis
September 29, 1985
University of Illinois at Urbana/Champaign

Professor Douglas has asked me to do two things: first, describe briefly the two-year calculus course at University High School in Urbana, Illinois (called "Uni"); and second, give the perspective of a mathematician who has become a full-time student of educational matters. This paper is in three sections: first, the Uni calculus course; second, a modern view of how information may be coded in the human mind; and third, differences to be found when one compares different courses in mathematics at various levels.

I. CALCULUS AT UNIVERSITY HIGH SCHOOL

Uni students are younger than those in neighboring public schools, often graduating from high school by age 16 or 17. By age 14 or 15 many have completed the study of typical "high school" mathematics, and consequently Uni has a two-year sequence in calculus, taught to juniors and seniors (who are typically 15 or 16 years old). The point of discussing student age so prominently is that Uni students are further ahead in mathematics curriculum than most U.S. students their age. How should one take advantage of this situation?

We have argued that one ought not to aim primarily for moving through the curriculum as quickly as possible, but can instead best serve these intellectually-gifted students by seeking a deeper-than-usual level of understanding, and also by trying to help them build up ingenuity and strategic-planning skill by emphasizing heuristics and methods of analyzing mathematical problems.

Although, hopefully, the students do not see the design principles that underlie the creation of the calculus course, it may be wise to state two or three of them here:

1. We want students to see mathematics as a reasonable response to a reasonable challenge (and not as an arbitrary collection of meaningless rituals).

2. We want the student's new challenges to grow out of their experience--which of course means that we must often begin by providing that experience.

3. In general, we want to be honest with students--let them know where difficulties lie, how we will deal with them, where special attention is needed, and where we will defer discussion until later.

4. We try to recognize explicitly a number of different goals. Learning key concepts is clearly a major goal; acquiring routine skills is an obvious goal; developing more subtle or "creative" analytical skills is a goal, but one that is less common in mathematics instruction in general; being able to read sections in the calculus textbook on your own is a goal (and, again, one that is

usually neglected in most courses); being able to use a
formula correctly even when you don't understand it fully
is a very common goal in many mathematics courses (and in
far too many engineering courses)--we give it very little
place in this course, but we do include it, explicitly.
After all, I may want to select, say, an air conditioner
for my home without learning much about insulating
materials, etc., by merely "putting numbers into a
formula". We also want students to be able to tackle
larger problems where one single problem may take several
days to solve, and may require looking up methods that one
does not know from memory (such as various numerical
analysis or successive approximation methods). Finally, we
want students to become skillful in relating mathematics to
real-world phenomena, especially in physics.

Let me give some examples of what the course itself
looks like:

1. The first task we have tackled in recent years is
computing the work done in compressing a spring. It is
clear that computing the work done by a force is easy
whenever the force is constant. In this case the force is
not constant. Suppose we _pretend_ it is constant--we find
that we can (easily) get bounds on the possible error from
this assumption. Suppose we use this assumption over only
half the distance--we find that the total error is smaller.
We can continue this strategy, assuming a constant force
over only 1/4 of the distance. We consider the

possibilities in general, when one has an increasing number
of possibly additive errors, but the individual errors are
getting smaller. In the present case, fortunately, we can
easily get bounds on the total error. We see that, given
any allowed error, we can make the actual error smaller
than that.

        Hence, we move on to consideration of the limit of an
infinite sequence.

        This analysis is re-interpreted geometrically as
finding the area of a triangle, and the method is applied
to finding areas under parabolas and other curves.

        In the course of doing this we are led to:

i)   Finite difference methods to conjecture summation
        formulas such as

        $$1^2 + 2^2 + \ldots + n^2 = \frac{n(n + 1)(2n + 1)}{6} \; ;$$

ii)   Mathematical induction for proving such formulas;

iii)   The use of the Law of Trichotomy and indirect
        proofs to show that the limit of .9, .99, .999,
        ... is exactly 1;

iv)   Consideration of the role of axioms in such
        proofs, and NOT merely intuition;

v)   Formulation for a correct definition of the limit
        of a sequence;

vi)   Proof that the limit of a sequence, if it exists,
        must be unique;

vii)   Intuitive formulations of the meaning of
        convergence of infinite sequences.

This is how the course begins. We would argue that it gives a reasonably representative view of what calculus is all about--a serious view, and one that we can build on for the rest of the two years.

2. As a second example, cf. our next treatment of sequences. We want the students to see why such a thing as a "sequence" is worth considering, so we deal with a few examples:

    i)   Inscribing "triangular sectors" in a circle, to approximate π;

   ii)   Considering $(1 + h)^{1/h}$ for h=1, .1, .01, ...;

 iii)   Approximating $\sqrt{2}$ by numbers of the form $N \times 10^{-R}$, where we look for the smallest number that is too large, and the largest number that is too small, thereby getting:

        2, 1.5, 1.42, ...

        1, 1.4, 1.41, ...

  iv)   Considering:

$$\frac{\sin \theta}{\theta}$$

        for θ = 1, .1, .01. ... radians.

In each case, the students program a programmable calculator or computer to produce a sequence of approximations. The students are asked to try to imagine all possible "pathologies" that might appear in sequences, such as:

        1, 0, 1, 0, 1, 0, ...

        1, .9, 1, .99, 1, .999, ...

        1, −1, 2, −2, 3, −3, ...

and so on.   It is left for students to try to formulate an appropriate definition of convergence; each candidate is checked against the "monsters" that have been collected, to see if it makes a correct delineation between "what ought to be called convergent" and "what ought to be called divergent".

3. Finally, as an example of a "larger problem that may take several days to solve", here is one: Recently a batter hit a home run completely out of the park in Detroit. How fast was the ball traveling when it left the bat?

Clearly, one has to begin with some estimates of the distances and heights involved. Clearly, also, one has to reinterpret the problem, turning it into one question, or several questions, for which reasonable answers can be found. (Students often, at first, confuse this with the problem of determining the angle of elevation of a cannon shot so as to maximize the distance traveled if muzzle velocity is fixed; they must think through how the height of the center field bleachers alters the nature of the problem.) Some of the relevant calculations are non-routine. (This course is described in more detail in Davis, 1984, and in Davis, Young, and McLoughlin, 1982.)

Two questions might be raised: (i) Is this course in any way different from most calculus courses?, and (ii) Is it relevant to typical needs in other settings? Our observations of other courses gives a strong affirmative answer to the first. To answer the second question requires more of an act of faith, but we would argue that the educational values we pursue ought indeed to be relevant in many other settings. (Various examples of work done by students in this course are published from time to time in the Journal of Mathematical Behavior.)

II. MENTAL REPRESENTATIONS OF MATHEMATICAL KNOWLEDGE

Two or three decades ago the typical psychological study of a "concept" dealt with what I would prefer to call a "decision rule"--for example, wooden blocks of different sizes, shapes, and colors might be used. Some would be "selected", others would be "rejected", and the subject was supposed to determine the rule that shaped these decisions (which might be something like: "either large and green, or else small and not blue"). I do not think this kind of study correctly captured what mathematicians usually think of as a concept.

I would argue that a mathematical "concept" is a much larger thing, and it contains important internal structure. The concept of "limit of a sequence" includes: (i) knowledge of specific examples that make clear why one is studying this kind of thing; (ii) knowledge of what the general goals are in studying such things; and some background notion of which sequences ought to be "useful" and which seem unpromising; (iii) knowledge of several definitions, at different levels of intuitive power or formal clarity; (iv) knowledge of approximations, and how the limit of a sequence is an essentially different kind of thing from a mere approximation (e.g., because of uniqueness); (v) knowledge of how to prove that the limit of a sum is the sum of the limits (when these exist), and other similar theorems; (vi) knowledge of how axioms are used in making these proofs; (vii) knowledge of how indirect proofs are used; and other similar matters.

Even if you feel that this includes too much, I suspect most mathematicians would agree that a "concept" is some kind of idea, some mental representation that has its own internal structure. One can speak of "the concept of derivative", or "the concept of indirect "proof", or "the concept of a vector space", or "the concept of an analytic function of a complex variable". Just as the concept of an airplane involves knowing parts (such as wings, landing gear, engines, etc.) and the relations to one another (plus the fact that it takes off, flies, and lands, and so on) something similar is the case with most mathematical concepts.

Whereas the earlier psychological studies did not represent this "internal structure" in a satisfactory way, recent work in cognitive science has done much better, as shown by the approaches of Minsky, Papert, Schank, Charniak, Lakoff, Simon, Fahlman, Rissland, and others.

What is relevant for our present concern with calculus seems to be perhaps five things:

1. Learning the key <u>concepts</u> of calculus is one of the main goals of studying the subject;

2. "Concepts", in the precision, subtlety, and abstraction of mathematical concepts, are not a common experience of human beings; most people most of the time get by with far less well-formed "concepts" as in (say) "the concept of 'war'", or "the concept of 'love'", or "the concept of a 'city'";

3. Building an adequate mental representation for an important mathematical idea is often hard work; most people do not seem to do it easily (and people often try to avoid it);

4. Alternatives--of a sort, anyhow--<u>do</u> exist, and people often resort to them. Some of the most common alternatives focus on <u>notation</u>, paying scant regard to what the symbols <u>mean</u>. (This seems to be true from the earliest arithmetic through contour integration in complex variables and the Einstein summation convention in tensor calculus.)

5. Finally, a major question facing us is whether the
    existence of these "alternatives" (or
    pseudo-alternatives) that seem to obviate the need
    to master concepts is in fact a welcome blessing or
    a harmful curse.

When one sees large numbers of inexperienced teaching
assistants presenting a "notational" calculus to vast
numbers of freshmen who may not be prepared to think about
mathematics, it can seem that only the existence of
non-conceptual "manipulative" notational alternatives
allows the whole enterprise to continue to operate.

But when one looks carefully at some individual
students, who seem never to suspect that symbols can have
meanings, the whole scene becomes far less satisfying. In
our own studies we have found this at every level from
beginning arithmetic through some upper-class university
courses in engineering, from a sixth-grader who wrote
more than one decimal point in a single numeral--as in
.3 + 4. = .7.-- with the symbol .7., whose size was unknown
to her (as, in fact, were the sizes of 4. and .3), to
calculus students who cannot solve the "what angle of
elevation maximizes the distance" problem because it never
occurs to them to consider y=0 as being in some way
relevant to the task at hand. These students show every
evidence of having suffered considerably from their pursuit
of a meaningless, ritualistic manipulation of symbols.
(Cf., e.g., Davis, 1985.)

III. TYPES OF COURSES

We have believed that we can distinguish a number of different approaches to the teaching of calculus:

1. By far the most common is a rather superficial course that focuses on <u>notation</u>, on routine problems, and on a few simple applications. Students learn to write derivatives in relatively simple cases, anti-derivatives in easy cases, something about the meaning of rates of change, etc. This is NOT necessarily an <u>easy</u> course, but the difficulty comes mainly from the rapid pace of moving through the material, and from an attempt to cover a large number of details without much focus on <u>main key ideas</u>.

In fact, one can defend such a course on several grounds:

i)   From the point of view of selecting students for advanced work, this course serves a useful screening role because of its "survival of the fittest" approach;

ii) Many people have passed through such courses and emerged as successful scientists, engineers, or mathematicians;

iii) If the course does not take the time to develop concepts and problem-analysis strategies in much depth, one can argue that that will come later, in courses in engineering and so on.

iv)   The most extreme argument might be based on the
      claim that a superficial introductory course may
      follow an unavoidable law of human
      learning--perhaps humans MUST acquire a
      superficial, gap-and-error-laden version of any
      knowledge before they can begin the process of
      refining and improving it.

(So as not to confuse the reader, let me say that I still
prefer an introductory calculus course that tries to
achieve much greater depth and to develop more
problem-attack skills. But it would be foolish to act as
if opposing arguments do not exist.)

To return to our list of alternative types of courses:

2.   Some mathematicians think that the main alternative to
a superficial calculus course is a course in real
variables. This is NOT what I would propose as an
alternative for beginning students.

3.   A common alternative to the "superficial" course is one
where some (more-or-less) correct proofs of theorems are
shown to students. As usually implemented, this approach
fails. Most students do not understand the proofs, do not
have a clear perception of the difficulties that must be
overcome, could not repeat the proofs, and surely cannot
devise original proofs of this type.

4. A viable alternative <u>does</u> begin to appear when we try
for what might be called "cognitive clarity". Key ideas
are presented very carefully and thoroughly, so that
difficulties are clearly perceived and so that students are
able to see how these difficulties are met and dealt with.
Some details are perhaps slighted, in the interest of
clarity and emphasis--but not to the extent that anything
serious is lost.

5. Further variations appear when one adds an extensive
consideration of problem-analysis heuristics.

6. One can also add "projects", perhaps undertaken by
several students working together, that may combine
modeling a real situation, writing computer programs, using
some numerical analysis or successive approximation
methods, etc.

7. The most important "additive" for me is what I think of
as cognitive (or even epistemological) clarity. This
involves being accurate about the nature of our goals and
our methods. If the question is to use axioms to achieve a
more precise foundation for something of which we have
intuitive or vague knowledge, that task should be made
clear, and where possible the students should be invited
inside the operation and should become participants.
Similarly for any other aspect of the work.

In particular, how <u>we</u> think about key concepts should be made clear, at least to some extent. How do <u>you</u> think of a <u>function</u>? A machine with inputs and outputs? A mapping? However you think of it, <u>that</u> should be made clear to students.

This is not a common approach, and I fear that I may be misunderstood. I don't mean something very personal and vague and complex--many good mathematicians and scientists of my acquaintance have some strikingly <u>simple</u> and <u>powerful</u> ways of thinking about key ideas. Our experience is that students, too, can find these ideas both powerful and simple, if we take some pains to let them in on the secret.

Bibliography

Davis, Robert B. <u>Learning Mathematics: The Cognitive Science Approach to Mathematics Education</u>. Norwood, NJ: Ablex Publishing Corporation, 1984.

Davis, Robert B. "Learning Mathematical Concepts: The Case of Lucy". Journal of Mathematical Behavior, vol. 4, no. 2 (1985), pp. 135-153.

Davis, Robert B. Young, Steven, and McLoughlin, Patrick "The Roles of 'Understanding' in the Learning of Mathematics". Part II of the Final Report of the National Science Foundation Grant NSF SED 79-12740, April, 1982. (Available from the University of Illinois at Urbana/Champaign, Curriculum Laboratory, Urbana, IL.)

THE LOGIC OF TEACHING CALCULUS

Susanna S. Epp
Department of Mathematical Sciences
DePaul University
Chicago, Illinois 60614

The Problems

Eight years ago my department instituted a course in mathematical reasoning to serve as a transition between calculus and higher-level math classes. We had found that students were entering our higher-level classes woefully unable to construct the most simple proofs or to figure out answers to easy abstract questions. The idea of the course was to give students a better chance for success in more advanced classes (1) by teaching the basic techniques of mathematical proof in such a way that students would learn to use them themselves, and (2) by spending an adequate amount of time on the rudiments of set theory, equivalence relations, and function properties rather than hurrying through these topics quickly as often happens at the beginning of advanced courses.

Over the next several years I had primary responsibility for developing the course. During this period I came to realize that many of my students' difficulties were much more profound than I had anticipated. Quite simply, my students and I spoke different languages. I would say "Of course, this follows from that" or "As you can see this means the same as that" and my students would look at me blankly.

Very few of my students had an intuitive feel for the equivalence between a statement and its contrapositive or realized

that a statement can be true and its converse false. Most
students did not understand what it means for an "if-then"
statement to be false, and many also were inconsistent about
taking negations of "and" and "or" statements. Large numbers used
the words "neither-nor" incorrectly, and hardly any interpreted
the phrases "only if" or " necessary and sufficient" according to
their definitions in logic. All aspects of the use of quantifiers
were poorly understood, especially the negation of quantified
statements and the intepretation of multiply-quantified
statements. Students neither were able to apply universal
statements in abstract settings to draw conclusions about
particular elements nor did they know what processes must be
followed to establish the truth of universally (or even
existentially) quantified statements. Specifically, the technique
of showing that something is true in general by showing that it is
true in a particular but arbitrarily chosen instance did not come
naturally to most of my students. Nor did many students
understand that to show the existence of an object with a certain
property, one should try to *find* the object.

The conclusions I came to through observing my students are
in substantial agreement with the results of systematic studies
made by modern cognitive psychologists. As the British
psychologist P. N. Johnson-Laird put it in 1975: "It has become a
truism that whatever formal logic may be, it is not a model of how
people make inferences." [5] A common estimate is that under 5%
of people use "correct" logic spontaneously. Even Piaget in his
later years came to modify his view that the development of formal

modes of thought was a natural occurrence at a certain stage of adolescence and acknowledged that his original work had been based on a "somewhat priviledged population." [8]

In a course such as mine, the consequences of such poor intuition for logic and language were devastating. For example, at one point in the course students were asked to prove that the sum of two rational numbers is rational. Consider what thought processes are involved in creating such a proof. Here is a partial list.

(1) One must understand, either consciously or subconsciously, that the statement is universal, that it says something about *all* pairs of rational numbers.

(2) One must realize that to prove this universal statement is true, one supposes one has two particular but arbitrarily chosen rational numbers and shows that *their* sum is rational. (That is, one must understand either consciously or subconsciously the method of proof using the "generic particular.")

(3) One must know both that if a number is rational then it can be expressed as a quotient of integers and also that if a number can be expressed as a quotient of integers then it is rational. (That is, one must understand how to use both directions of a definition: the "if" and the "only if." Also it is helpful to associate a vision of a blurry fraction with the term rational.)

(4) One must understand the rule for adding fractions as an abstract universal truth that can be applied in an general algebraic setting.

(5) Since virtually every step in the proof is a conclusion of a syllogism, one must understand how conclusions follow in syllogistic reasoning by applying universally applicable facts to particular instances.

At another point in the course, students were asked to prove by contradiction that the negative of an irrational number is irrational. To succeed at this task, one must realize that if the given statement is false then there is an irrational number whose negative is rational. (That is, one must be aware at some level of consciousness that the negation of a universal statement is existential.) Also, of course, one needs a sense for the logical flow of proof by contradiction.

At still another point in the course, students were asked to prove that the composition of one-to-one functions is one-to-one. To construct a proof of this statement, a really sophisticated ability to instantiate is necessary. One must understand that when a function f is one-to-one, the statement "if $f(x_1)=f(x_2)$ then $x_1=x_2$" holds for *all* x_1 and x_2 *even when* x_1 happens to be called $g(x_1)$ and x_2 happens to be called $g(x_2)$.

As noted, the above understandings need not be at a conscious level. Lots of working mathematicians have never studied formal logic and get along just fine. Unfortunately, that is part of the problem. On the one hand we have the professor for whom formal reasoning is second nature and who is usually not even consciously aware of the formal logical components of mathematically correct arguments. And on the other hand we have a mass of students for most of whom hardly any of the logical component elements of arguments are understood on an intuitive level. The lack of insight of the professor to students' problems with logic and language are manifested in many ways. For example, it is common nowadays to omit the words "only if" in formal definitions. Supposedly this is in the interest of "simplicity." In fact, in my experience, if one hopes to impart to students a useful working knowledge of a definition, it is not only necessary to state the definition in "if and only if" form but also to state the "if" and the "only if" directions as separate sentences and to emphasize the universal character of each direction. For example, in giving the definition of rational number one needs to explain both that whenever a quantity in a discussion is known to be rational then it must be a quotient of two integers and also that whenever a number is known to be a quotient of two integers then one can infer that it is rational. Nor is it sufficient to state that an irrational number is one that is not rational. One must go on to explain that this means the number cannot be expressed as a quotient of any two integers.

Similarly, it is common in mathematical writing to leave out or to veil the presence of universal and existential quantifiers. As Alan Bundy states when introducing the concept of quantifier in his book *The Computer Modelling of Mathematical Reasoning* [2]: "Variables in mathematical expressions often have ambiguous status, whose resolution depends on the context." He then compares the x's in the two sentences

$$(x-y)(x+y) = x^2-y^2$$

and

$$\text{Solve } x^2+2x+1 = 0 \text{ for } x$$

explaining that in the first case the universal usage is
ordinarily intended while in the second case the existential usage
is meant. He next gives an example of a single sentence in which
the two usages are combined:

$$\text{Solve } ax^2+bx+c = 0 \text{ for } x.$$

Now to a mathematician this problem is perfectly clear. But to a
high school algebra student the status of the variables may seem
mysterious indeed.

Implications for the Teaching of Calculus

The implications of these observations for the teaching of
calculus are profound. Calculus has so many definitions, so many
theorems, so many applications, so much notation, so much "abuse
of language" (as the French call it), so much logical complexity,
so much abstraction. Consider a student who does not even know
that to prove a sum of two rational numbers is rational one starts
with two arbitrarily chosen rational numbers. How can such a
student begin to understand even the most "intuitive" explanation
that the limit of the sum of any two functions (which have limits)
is the sum of their limits? Not to mention the fundamental
theorem of calculus!

It seems that most mathematics professors at most colleges
and universities are aware of an intellectual gulf between
themselves and their students. Last year at the Joint Mathematics
Meetings the Association for Women in Mathematics sponsored an
panel which featured five mathematicians who had left academia for
employment in business or industry. One panelist after another
spoke of being disillusioned with the quality of student they had
had to teach during their periods as academics. Maria Klawe

(Discrete Mathematics Manager at the IBM San Jose Research Laboratory) was especially eloquent as she spoke of her despair and frustration at the thought that all the years she had spent in graduate school developing her capacity for abstract thought would be wasted in an effort to teach students who would not and could never learn college mathematics. While it was not too surprising that the panelists spoke so disparagingly (they had after all chosen to leave classroom teaching), what was surprising was the loudly sympathetic reaction of the large audience.

Professors react to the gulf between themselves and their students in different ways. One reaction is to ignore it, to state the definitions and prove the theorems as if the students would understand them and were mature enough to be able to derive simple consequences (such as problem solutions) on their own. This approach is often associated with the calculus instruction of 20 to 30 years ago and is fondly remembered by many mathematicians. (It is bitterly remembered by many physicists and engineers.) Another response, widely adopted today, is to expose the basic concepts of calculus at a moderately high level, emphasizing intuition, but focus primarily on skills, and only test students on their ability to perform certain mechanical computations in response to certain verbal cues. In this approach the vast majority of students indulge their professors by listening quietly to their often inspired and beautifully intuitive explanations and the professors repay their students' courtesy by making their explanations brief, spending lots of time demonstrating procedures to solve rote problems, and never asking students to do anything on an exam that requires genuine knowledge of concepts.

Of course, there are good reasons for giving attention to the mechanical aspects of calculus. The best is based on the sound pedagogical promise that understanding in mathematics comes in pieces. Often, learning to use certain techniques mechanically helps one progress to conceptual understanding. But this approach becomes perverted if in practice conceptual understanding is indefinitely postponed. Another benefit of emphasizing mechanics is that such activities as practice in formal differentiation and integration improve students' pattern recognition skills. A possible third argument in favor of an emphasis on calculus mechanics is to prepare students for courses in physics and economics and engineering. But I won't make this argument. For too many years I have heard complaints from my colleagues in other departments about the mathematical knowledge of the students we send them. Invariably, the "simple" examples they give as evidence that our students can't perform involve being able to *think*, not just compute on cue.

The fact is that the state of most students' conceptual knowledge of mathematics after they have taken calculus is abysmal. The most dramatic formal studies on this subject have been done by John Clement, Jack Lochhead, and others in the Cognitive Development Project at the University of Massachusetts at Amherst. They found that a large majority of calculus and post-calculus students tested at universities throughout the country could not set up or even correctly interpret simple proportionality equations. In summarizing the results of their many experiments, Lochhead wrote: "many college students are not

facile at reading or writing simple algebraic equations" and at a
deeper level "students seem to lack any well defined notion of
variable or of function." [6] Currently Hadas Rin is studying
student difficulties with calculus by examining their spontaneous
written questions. Among her findings are the misuse of course
vocabulary by students (for example, "How do you find the tangent
to the slope?" or "Any number has no derivative"), the inability
to instantiate "known" theorems in new situations (for example,
asking for "the rule" to differentiate a function of the form
$\frac{f \cdot g}{h}$), and lack of understanding of definitions, not just of
sophisticated concepts such as limit but also of more fundamental
ones such as secant and tangent. [9]

 To me it seems incontrovertible that the primary aim of
calculus instruction should be the development of conceptual as
opposed to purely mechanical understanding. In this computer age,
software packages are now or soon will be available to perform any
standard calculus computation including taking symbolic
derivatives and integrals and even testing series for convergence.
People using these packages need some computational facility
themselves (just as experience doing arithmetic by hand is needed
for a person to make best use of a calculator). But the main
requirement to use calculus packages effectively is firm
conceptual understanding of the subject matter. With computers to
take care of mechanical details, the premium is on the abilities
to abstract, to infer, and to translate back and forth between
formal mathematics and real world problems. Yet these are
abilities of the highest order, normally associated with a small
number of students of exceptional talent.

Suggestions

Never in history have mathematicians been called upon to teach so much mathematics to so many students. Under these circumstances, it should not be surprising that new pedagogical methods may be necessary. The shortcuts and gaps that can be followed and filled in by students of unusual ability may not be negotiable for those less fortunate. To a much greater extent than is currently the case, there is a need to respond to students' lack of sophistication, not by giving up but by helping them.

One possibility is to modify precalculus courses to make them include additional work to increase students' logical maturity. I see this as a potential benefit of the movement to introduce discrete mathematics early in students' undergraduate careers. Within the context of a course in calculus, I would suggest the following measures. Some (perhaps all) may be controversial. I have found all of them useful.

(1) State logically complex sentences (such as the definition of limit) and pose problems in a variety of equivalent ways. Left to themselves, students usually do not turn concepts over in their minds to view them from many angles. For instance, one could ask students to

"Describe the values of the expression $\frac{(x+1)}{(x-1)^2}$ when

x takes values very close to 1"

as well as to

"Find $\lim_{x \to 1} \frac{(x+1)}{(x-1)^2}$."

(2)   When lecturing, write more or less in complete sentences. When the words "if-then" or "for all x in the interval [a,b]" are not written out, they will not appear in students' notes, nor will they be supplied mentally.

(3) Make an effort to clarify statements whose quantification is implicit. For example, the implicit quantification of the phrase "solve the equation" goes hand-in-hand with a mechanical approach based on formal symbol manipulation rather than a conceptual approach based on studying numbers and their properties. Leon Henkin suggests that teachers of beginning algebra students avoid using phrases like "Solve the equation $2x+3 = 0$" and instead say "Find a number x such that $2x+3 = 0$." And instead of "Solve $ax+b = 0$" he suggests "If a and b are numbers and $a \neq 0$, find a number x such that $ax+b = 0$." [4] Or, instead of asking students to solve equations, one could ask them questions like "Are there any real numbers such that $x^2-3x+2 = 0$?" or "such that $x^2-x+1 = 0$?" or "such that $\sqrt{x+1} + \sqrt{2x+1} = 2$?" College calculus students would also benefit from occasionally having problems phrased in these ways.

(4)   Avoid unnecessary notation and terminology. For most students, each mathematical term and symbol is a hurdle to be crossed. Let's not put any more in their way than we have to.

(5)   Try to avoid notational and linguistic "abuses" as much as possible. In the long run, it is worth the extra effort to say "Let f be the function defined by the rule..." rather than "Let $f(x)$ be the function..."

(6)   Frequently clarify lines of argument by explaining the underlying logic.

(7)   Make students memorize precisely-worded definitions and perhaps theorem statements also.   Memorization is greatly underrated as a pedagogical tool.   At the least, memorization of a definition or theorem forces students to read it carefully; at best, it encourages them to understand it (since it is easier to memorize something intelligible than gibberish).   Also memorization of precise language gives students experience in using it and makes it necessary for them to pay attention to words like "if" and "then" that they might otherwise ignore.

(8)   Develop or seek out problems to act as cognitive bridges to abstract understanding.   I have found, for example, that students are fairly capable of understanding concepts in purely geometric terms.   They do not seem to have problems learning to distinguish between the graphs of continuous and discontinuous functions or between concave up and concave down.   The difficulties arise when analysis is added to the picture.   One reason is that, in my experience at least, beginning calculus students do not know the abstract definition of graph of a function.   They can plot and connect points for a function given by a specific formula, but they do not know that for a general function f, $f(x)$ is the height to the graph of f at x.   Now since most calculus explanations are given in terms of "generic" functions f and "generic" points x, this seriously inhibits students' ability to follow text and classroom explanations.

To counteract this difficulty, I would suggest adding problems of the following type to the usual collection on graphing.

Let f and g be function defined for all real numbers.

(a) Suppose f(4)=9. What point must lie on the graph of f?

(b) Suppose the point (-1,2) lies on the graph of f. What can
     be inferred about f?

(c) Suppose the point (3,g(3)) lies on the graph of f. What
     can be inferred about the relation between f and g?

(d) Suppose the graphs of f and g have a point in common, say
     (x_0,y_0). What can be inferred about the relation between
     $f(x_0)$ and $g(x_0)$?

     Later, just prior to the introduction of the analytic

definition of the slope of the tangent line, one would assign

exercises such as these.

1. Let f be the function whose graph is given below.

     (a) Label the points (2,f(2)) and (4,f(4)) on the graph and
          draw the secant line through these two points.

     (b) Find an expression for the slope of the secant line
          through the points (2,f(2)) and (4,f(4)).

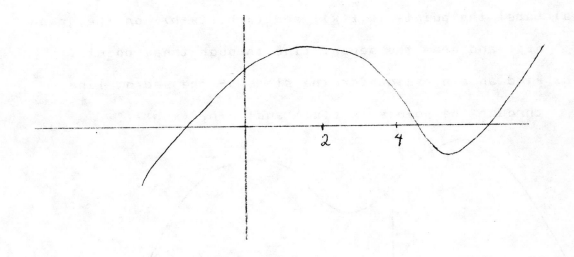

2. Let f be the function whose graph is indicated below and
suppose h is a (small) positive number.

   (a) Label the points (3,f(3)) and (3+h,f(3+h)) and draw the
       secant line through these points.

   (b) Find an expression for the slope of the secant line
       through the points (3,f(3)) and (3+h,f(3+h)).

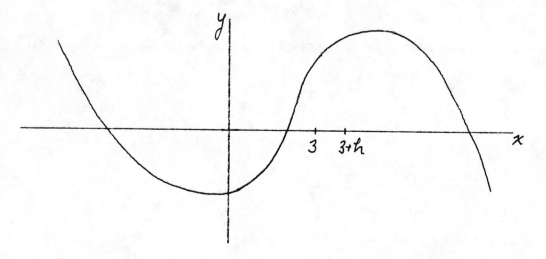

3. Let f be the function whose graph is indicated below.
Suppose also that x is a number and h is a positive number and
f takes values at x and x+h.

   (a) Label the points (x,f(x)) and (x+h,f(x+h)) on the graph
       of f and draw the secant line through these points.

   (b) Find an expression for the slope of the secant line
       through the points (x,f(x)) and (x+h,f(x+h)).

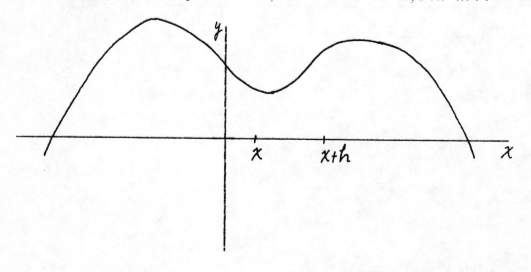

4.   Let x be the number represented by the labeled point on the number line below and suppose h is a *negative* number.   Indicate and label a reasonable choice of point to represent the number x+h.

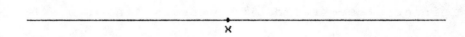

(9) Do not be satisfied with narrow understanding of abstract principles.   Throughout their mathematical careers students have great difficulty learning universal facts in their generality. (Logically speaking, they have difficulty instantiating universal statements over the full extent of their domains.)   Much mathematics instruction takes these problems into account.   For example, precalculus texts often have exercises that are graded with problems like "Factor x^2+5x+4" in the "A" set, "Factor $3x^2$-14x+8" in the "B" set and "Factor $24x^2$-31xy-15y^2" in the "C" set. [1]   In calculus, also, concepts can be understood at "A," "B," and "C" levels of generality.   For example, in a study of the chain rule an "A level" problem would be

"Find $\frac{dy}{dx}$ for y = $(3x+2)^2$,"

a "B level" problem would be

"Find $\frac{dy}{dx}$ for y = $\sqrt{\sin 2x}$,"

and a "C level" problem would be

"Assume f is a differentiable function.   Let y = $[f(x)]^4$ and suppose f'(1) = 5 and $\frac{dy}{dx}\Big|_{x=1}$ = -160.   Find f(1).   [3]

It is ironic that the same students who, as second graders, were judged needful of spending a full year on two-place addition

before moving to the complexity of the three-place case are as
18-year olds expected to generalize in one homework assignment
from "A level" chain rule problems to "C level" ones. Perhaps
problems of increasing difficulty could be assigned and discussed
over a period of several days, concurrently with other topics if
necessary, to allow time for the abstraction process to occur.

(10) Include questions that test conceptual understanding on
homework and on exams. In [7] Jean Pedersen and Peter Ross make
some excellent suggestions of such problems, which test
understanding both of geometric and analytic aspects of concepts.

(11) Take responsibility for all aspects of students'
mathematical development. We do not help our students when we
ignore "mere" algebra mistakes just because algebra is not the
subject of our course.

(12) Give students opportunities to speak and write using the
course vocabulary.

(a) Insist that students give complete, coherent answers to
questions on exams. No favor is done students' intellectual
development by giving full credit for a few scribbles.

(b) Have students present solutions to problems at the board
occasionally, and insist that they explain their work aloud to the
rest of the class. You may be appalled by their misuse of
language, but don't despair. Just correct the worst mistakes
courteously and find something to praise. Most people do not
learn to speak a foreign language by attending lectures and doing
grammar exercises. They have to make fools of themselves by

talking out loud. The same goes for learning the language of mathematics.

(c) Occasionally break classes up into small groups for collective problem-solving experiences. Even if this is done only once or twice a semester, it can have impact on students' ability to put their mathematical thoughts into words. Collaboration also encourages students to explore new ideas more boldly than they would on their own.

(d) Every once in a while restrain the impulse to give the answer to a question on the homework as soon as it is asked. Instead, open the question up for class discussion. (This works only for carefully selected problems.) In one of the liveliest classes I have ever taught, a student asked me to answer the following homework question.

"Determine whether the following statement is true, and support your answer by giving a proof or a counterexample:

If a, b, and c are integers and $a \nmid b$ and $a \nmid c$ then $a \nmid b \cdot c$." Instead of complying, I told the class to imagine they were the mathematical research division of a large company and had been given the job, as a group, of figuring out the answer to this question. Important decisions depended on the answer and the company was counting on its being correct. In the discussion that followed, I acted as moderator. I guided, but I did not reveal the answer. At one point the board contained two false proofs, a false counterexample, and a true counterexample. (I am happy to report that, overwhelmingly, the students were convinced by the true counterexample when it finally appeared.) During the discussion almost all the logic we had covered to that point in the course was reviewed and its importance to the determination of the mathematical truth of the situation was apparent. There was also a lively follow-up discussion on the nature of mathematical discovery.

Conclusion

If all the suggestions made above were incorporated into the teaching of calculus, it would probably be impossible to cover as many topics as is now standard. The question is: What is the trade off? If it comes to a choice, will we settle for superficial knowledge of a lot or deeper understanding of less? Perhaps less is more.

Do I think that by following these suggestions a new breed of mathematical super-students will be created? Certainly not. But I do think there are many students "out there" of reasonable mathematical aptitude for whom mathematics is more mysterious than it needs to be. These are the students we can affect with better pedagogy. I think it is worth trying.

References

[1] Barnett, Raymond A., *Intermediate Algebra: Structure and Use*, Second Edition. New York: McGraw-Hill, 1980.

[2] Bundy, Alan, *The Computer Modelling of Mathematical Reasoning*. London: Academic Press, 1983.

[3] Fraleigh, John B., *Calculus with Analytic Geometry*, Second Edition. Reading, Massachusetts: Addison-Wesley, 1985.

[4] Henkin, Leon, "Linguistic Aspects of Mathematical Education." In *Learning and the Nature of Mathematics* edited by William E. Lamon, pp. 211-218. Chicago: Science Research Associates, 1972.

[5] Johnson-Laird, P. N., "Models of Deduction." In *Reasoning:
     Representation and Process* edited by Rachel J. Falmagne, pp.
     7-54. Hillsdale, New Jersey: Lawrence Erlbaum Associates,
     1975.

[6] Lochhead, Jack, "The Mathematical Needs of Students in the
     Physical Sciences." In *The Future of College Mathematics*
     edited by Anthony Ralston and Gail S. Young, pp. 55-70. New
     York: Springer-Verlag, 1983.

[7] Pedersen, Jean and Ross, Peter, "Testing Understanding and
     Understanding Testing," *Coll. Math. J.*, 16 (1985), 178-185.

[8] Piaget, Jean, "Intellectual Evolution from Adolescence to
     Adulthood," *Human Develop.*, 15 (1972), 1-12.

[9] Rin, Hadas, private communication, September 1985.

Calculus as a General Education Requirement

prepared for the Sloan Conference on Calculus
Tulane University, January 2-6, 1986
March 1, 1986 Draft

John W. Kenelly
Clemson University
Clemson, South Carolina 29634-1907

A liberal education develops one's wide spectrum of intellectual capabilities, and mathematics has traditionally played a major role in general education. Why not today? More precisely, why does everyone readily admit to a critical need for the study of mathematics on the one hand and, on the other hand, avoid it at all possible cost? The author contends that a sizeable part of the answer lies in the *education vs training* debate, and the mathematics community has let its courses, especially its calculus courses, become techniques-and-tools training courses -- thus, violating its general educational role.

Historically, calculus has been the end course in the fully educated individual's liberal education, and it suffers most from a dilution of mathematics requirements. In this paper, we hope to reestablish the creditable role of calculus in a liberal education and present a course philosophy that will maintain its position. Thus, our successive tasks will be to defend the course's role, critique the faults in the current presentations, and outline possible corrections.

1. General educational requirements and calculus.

Initially, let us examine the theme that mathematics deals with the *subtle* properties of *simple* systems. This contrasts quite sharply with the companion observation that the sciences deal with *simple* properties in *complex* systems. A linguist would note this difference immediately from the languages used by mathematicians as opposed to other scientists. In the case of mathematics, the terms are often very simple -- sets, groups, functions -- and yet the meanings are most elaborate in their richness and subtlety. In distinct contrast, scientists use highly complex terms with underlying simple meanings. The "cultures" are as marked by their languages as the Hawaiian culture is by its lack of a word for *weather* and the Eskimo by its hundred different words for *cold*.

If *subtle studies of simple systems* is the basic fiber of mathematics, then what are the general educational and course structural implications? A common study pattern error will illuminate some of these elements. To wit, Johnny has two exams: one in biology and the other in mathematics. He studies the complex vocabulary of his biology course and he gets a rather *simple* understanding of the *complex* terms. He takes the biology examination and he applies his cursory understanding of the complex terms to some standard and direct applications. He scores very well. Then he walks across the hall to his mathematics examination with a *simple* understanding of all the mathematical terms. Since the terms are common words and typically used in many other contexts, he spends little time on his mathematics test -- after all he had to bone up on all the unusual spellings in biology! The test requires -- as it should -- an analysis of some of the *subtle* properties of the simple terms and Johnny blows it! His parents and many others complain, and the mathematics instructor is reprimanded by the administration for ineffective teaching. As a result, the future mathematics tests will tend to be *simple* applications in very standard settings -- after all *mathematics is a basic skill*. The end result is "mimicry mathematics" -- the teacher works in class and then tests essentially the exact same problems and techniques. As a result, Johnny gets a good mathematics grade, but he does not develop the mental processes that are needed in thought provoking situations, and a less well-educated individual is the product.

In this context, calculus instruction has drifted into a similar pattern; the current "show and tell" practices violate the course's basic role. What is that role? Calculus provides answers to the very simple question, "How do things change?" The answers are rich and more subtle than anyone can imagine. Civilization has benefited immensely by the search for answers to this question, and each individual will be a better educated person for having participated in parts of this search. It is this development of the ability to make intellectual inquiries and synthesize one's observations in the dynamics of change that is calculus' strength and natural position. In almost the same manner that a philosophy student never expects to answer the question, "What is truth," the mathematics student should not be after pat answers and routine techniques. Both the mathematics student and the philosophy student should see the *journey* as their reward, and their own personal understanding of the nature of thoughts and things as their real goal.

The history of great ideas is a basic component in one's general education, and calculus is truly one of the great intellectual achievements of mankind. The successive discoveries of Descartes, Newton and Cauchy should be studied for their beauty and rich history alone. Through the historical review of intellectual ideas, general history courses claim to give students a valuable perspective on contemporary events. Surely there are ample problems in today's economic crises and overpowering technological changes to warrant a request that educated people understand the history of how mankind learned to quantify and analyze change.

A historical prespective also gives us a better understanding of the essential role of the calculus in everyone's general education. Since the laws of nature are differential equations, it is obvious that calculus and the age of modern science had to arrive in that order. The discovery of calculus introduced and opened the way for scientific understanding. In a like manner, an individual is educationally unable to fully function in a scientifically dominated world without an ability to understand the language of scientific relations.

Since *change* is a fundamental element in every conceivable human activity, a correctly structured calculus course should have universal importance and appeal. The mathematics community has been a willing participant in the public sham that calculus is primarily for *science and engineering* students, and it is past time for us to stand up and say that the study of change is essential to everyone's intellectual development! Many foreign cultures see the study of mathematics as much a part of one's general development as, say, literature, and students in foreign countries continually study mathematics. Strangely, though, in the United States middle level mathematics is reserved for a selected clientele, and the public does not see mathematics as a universal requirement after the initial high school years. Could it be that we have made calculus into a course for the few and missed its basic nature by our own tunnel vision?

If calculus is to concentrate on the study of change -- *how do things change*? They change as a result of something internal, something external, or something that can be "nothing at all." In the respective cases, one gets compounded growth and decay, the universal oscillatory cycles of nature, and constant change. In the latter case, nothing is being added to effect the change and, along with everything else, the change doesn't change either!

Change inherently suggests *connections* and that requires that we digress for a moment to discuss functions. Maybe the harm starts when the beginning algebra teachers fail to give the students a conceptual understanding of variables. For many students, variables are simply letter symbols used in manipulative practice exercises. As a result, functions are -- as the students have been carefully coached to say -- ordered pairs of these things. They miss completely the idea that functions capture all the spirit and essence of connections and interdependencies. Functions are, in one sense, simple connections; in a larger sense, they embrace all the subtle elements of input and output, control and observations, and cause and effects. They model the world itself. But most students miss this and see mathematics as tools and cold abstractions that have no relevance or meaning. Accordingly, the dynamics of these connections, i.e., calculus, is an equally meaningless exercise in symbol manipulation, hardly a course that one could justify as being an essential part of each student's general education requirements and surely a specialty course that is reserved for people who manipulate symbols.

Let's take a very elementary result from calculus and, through it, give an example of how the course could contribute to one's liberal education. If y = sin x, then y" = -sin x. In this one simple result we can note, and the liberally educated person should be able to note, the basic character and patterns of nature's fundamental cycle. In this case, the cycle could be one from business, or one from science, or one from history. The result is the same -- a cycle is driven by forces that are exactly opposite from its state. That is, prices are the lowest when supplies are the highest, distortion peaks when the resistance is the weakest, and social outcries are their loudest at the beginning of the end. In even simpler terms, people tend to scream on a roller coaster when the falling is essentially over. And more importantly, in the ebb and flow of economic cycles, leaders would be well advised to time their corrective actions by the shifts in the trends and not by the vocal levels of the populations. This kind of insight into the character of natural cycles should be part of everyone's basic education.

Countercyclical investors understand the natural forces in the stock market and they wisely "go against the crowd" in their buy-and-sell activities. Data reported on the television program, "Wall Street Week," on October 18, 1985, supports the wisdom of the countercyclical approach [4]. The Investors Intelligence Poll of Market Advisors noted that *popular opinion* may be the "best" indicator of the "worst" action possible.

Date	Dow Average	Advisors Predicting		Future Changes	
		Decline	Increase	in Dow	Time interval
Aug. 2, 1963	689.38	91.4%		+250	in next 21 months
Jan. 12, 1973	1047.49		62.6%	-470	in next 23 months
Dec. 13, 1974	577.60	63.5%		+425	in next 14 months
Jan. 14, 1977	983.18		78.8%	-235	in next 14 months
Aug. 13, 1982	784.34	65.7%		+500	in next 15 months
Oct. 18, 1985	1368.84	61.9%		???	????????

[Added note: By February 27, 1986 the market had increased 345 points to a high of 1713.99]

Calculus is a conceptual subject that gives special insight into the behavioral fundamentals of other processes as well. If you want to capture the essence of a process, then note its operating state, pay close attention to changes in that state, and most notably, try to judge the nature of the changes in that change. If you do this, then you will have captured *position, direction* and *trend* -- the key elements in the behavior of any system. A student in calculus would readily recognize these as the function, its derivative and its concavity. But how many of our typical calculus students would characterize functions and their first and second derivatives as the items that quantify the essential behavior of any process?

Why is it that verbal descriptions seldom go past the first and second order effects? Assuredly, evening television newscasters talk at great length about increasing and decreasing rates of inflationary increases, and business publications are loaded with charts that plot values of the percent change. But you would be hard-pressed even to name, in nonmathematical terms, anything that describes third derivative effects. It might be that human insight is limited by traditional discussions that seldom go past first and second order effects. And it might be that our inner ear has an accelerometer, never physically sensing anything beyond acceleration, and our vocabulary shows it. This is the character of how we discuss change, and this analysis should be a part of everyone's development. Later observations that physics deals with velocity and acceleration; statistics deals with means and standard deviations; and projective geometry deals with segment ratios and cross ratios (ratios of ratios) will clearly be natural instances of first and second order effects being emphasized. Educated people would have discussed first and second order effects in their calculus courses and would have seen them to be sufficient; thus, precluding any need for the inclusion of third and higher order effect in any of their other studies.

For every *yin* there is a *yang,* and for every study of change there should be a study of "non-change." This is the section of calculus that fails the most in its attempt to reach its general educational goals. Through the years, integration as area and volume have become a means in themselves. Anyone who has taught a senior level course in probability theory is all too familiar with the long struggle that it takes to teach probability and integration theory to students who cannot conceptualize the integral as anything but areas and volumes. Not only are we failing in the general education of students, we are failing our own educational interests as well!

In a course that develops the measurement of change, it would be most natural to ask for techniques that quantify the removal of change. The topic could easily be introduced by a review of the ideas of averages, i.e., the measure of a collection of events when the changes or variations in values are "removed." Students would readily note that you "add them up and divide by n." What more is the Riemann Sum? Maybe the initial examples and problems would have to be carefully picked to be over unit intervals, but that is a trivial organizational detail. The mean value theorem for integrals is now perfectly natural -- "somehow, somewhere, everything has an average." And all the mystery disappears from all those little rectangles being added; the volume is naturally the average cross section times length.

Later, weighted averages can be discussed and the integral extended to even more powerful ways of measuring the removal of variability and change. This suggests that probability, i.e., the quantification of chance, is a natural area to use as a replacement for substantial parts of the sections on integration techniques. Once the integral has been introduced as an average and Riemann Sums explained in terms of many ways to use averages, e.g., heights, cross sections, pressures; then numerical values for definite integrals would be routinely found in many varied contexts. Weighted averages would then be a natural extension and probability could be a rich addition to the course. Microcomputers would permit the easy calculation of the numerical values of the definite integrals and class and homework time would be devoted to the analysis of the ideas. The author knows that this proposal calls for everyone to raise the difficult questions of numerical convergence and the like. But careful organization of the material would avoid such convergence pitfalls. We need only to look at the current textbooks and the special care that we use to select surface area problems in order to see the selectivities that we will use in the name of good teaching. It has been said that there are only 16 workable surface area problems in elementary calculus -- 10 in the textbooks and 6 saved by the teachers for the test problems.

A general education calculus course should <u>not</u> "return to the good ole days." However, there are parts of the old calculus course that should be recaptured. (By the way, all the students that *used to be* -- weren't!) In some of the old and strangely brief texts, one will find reams of different and varied applications of Riemann Sums. Admittedly, convergence problems and the like are given short shrift, and some of the sections reduce to the casual introduction of formulas. But the students surely saw calculus and infinite sums as widely applicable techniques to be used in a rich collection of varied problems. Contrast that with today's typical 200-page treatment of volumes of revolutions and next to nothing else. Is it any wonder that our students think that the volumes are the ends in themselves? And, correspondingly, do you really think that the so-called level of rigor in these new "theoretically correct" treatments could handle anything other than the carefully controlled environment that the author selects?

The proposed course, with an emphasis on averages and "non-change," would prepare the students for a lifetime of dealing with averages, moving averages, quantified chance, and approximations. They might not be able to find the volume of a molded object of revolution, but I believe that their thirty-nine dollar calculator with its definite integral button could.

2. Defect in the current instructional approach to calculus.

In a simple summary, today's calculus essentially is an intense collection of mechanical manipulations of polynomials, with techniques and tools that students perceive to be ends in themselves. That premise is the profile of what the paper will now use as an outline to critique the current course.

The current calculus course is still thought of on campus as a bear, and along with organic chemistry, it keeps the world of science graduates pure. The pre-war German engineering schools had a better student selection technique and the screening process did not increase teaching loads. There, you had to form, with hand files, an almost perfect cube from a hunk of pig iron. The Herr Professor could easily use a caliper to check your "admissions test" results in a moment and there was a minimum amount of faculty distraction. The advanced classrooms were reserved for the *dedicated* few [3]. In many ways we see the same elements of artificial barriers in today's "show and tell and work like hell" calculus courses.

Most calculus examinations emphasize the manipulative skills that are stressed in calculus, and the students learn "how to play the game." The result is a collection of graduates who have skills that start to approach the abilities of some of the symbolic manipulators on today's microcomputers [2]. If bench engineers can find definite integrals on their pocket calculators as easily as they can find square roots, then we can hardly sell calculus as a course that develops "needed" manipulation techniques.

In most of today's calculus texts, one needs to get past the first two hundred pages and, after that, past the first twenty problems in each exercise group before one finds questions that involve anything other than polynomials or, in the later case, expressions that are essentially polynomials in disguise. Students who are taught that calculus is the mathematical study of motion, would get the impression from these textbooks that things move around exclusively on polynomials. This is hardly the case, and this becomes clear when one realizes that internal forces create exponential and logarithmic changes and external forces create cyclical or circular function responses. The author feels that the use of polynomial expressions, as an ease of entry tool to introduce topics, has been expanded and expanded to the point that the study of polynomials dominates the typical course, when in fact the basic nature of motion suggests that the emphasis should be on the transcendental functions.

We have already mentioned that students clearly perceive the course as being important for tools and techniques. Even though this student feeling may be incorrect, and I doubt that it is, the perception is still a fact with which the calculus instructors must deal.

In a critique of the problems in a calculus course, the current textbooks are always questioned -- and the publishers get off the hook with the accurate observation that they "publish what the marketplace wants." That may be the case, but we all need to examine carefully today's collegiate publishing and marketing process. There is a disturbing trend that introductory college-level texts are starting to be developed with the same procedures as the ones used for high school books. That is, a good marketing formula is assembled and word craftsmen are employed to write the successful outline. The texts are sold on the basis of "early trig" or "late exponentials" and not on the basis of mathematical merits. Departmental selection committees look through the "new models" to see that they match the formula specification required by the service audiences. For example, physics professors insist that so and so be covered in the first semester and engineering professors expect . . . . After that, particular parts are scanned and each person's favorite treatment of a particular topic must stand the test. Needless to say, the publishers have anticipated this and every possible variation is included. Why else would you find in one of today's successful texts the (distasteful) cancelling of formal differentials on one page and a rather correct theoretical discussion of differentiability on the next? Pity the poor student when gutter-level comments are routinely mixed with strange excursions into high levels of rigor. What we need is a mathematician to write a coherent and honest approach to the subject, but I expect such a text would not sell! And that fault is clearly the responsibility of the mathematics faculties.

3. Corrections that should be considered.

One's own examinations are the most readily available starting place for an individual teacher to begin to effect the essential changes that the course so desparately needs. We all know that the mathematical community has called for basic changes in the calculus [1]. And we all know that the commercial publishers are "staying awake nights" to redesign the texts for better market penetration! And surely there is not a single mathematics department that fails to have a committee charged with the consideration and review of the faculty's approach to calculus instruction. But, in the meantime, the problems are here and now, and the individual need not wait. Corrections can start with your own examinations.

Most calculus tests are time trials. The instructors have the misguided feeling that equal credit problems should be equally difficult. As a result, the typical test questions are all of the same level and length. If time is called at the end of the class, and this is the norm, then the test is very likely to be what the test specialists would call a "speed test." Instead, we need to be giving a "power test." That is, the questions should be of graduated difficulty, and they should proceed from an almost trivial question to one that is a special challenge. Ample time should be allowed and, if necessary, fewer but more conceptual and non-routine questions should be used. A good analogy is supplied by a hurdles race. If all the hurdles are equally spaced and of the same height, then the runners are being measured in time against each other. If a runner can clear any one of the jumps, then given sufficient time, he or she can clear them all. A better test of hurdle strength would be provided by a sequence of several hurdles of increasing heights. Then, runners would be measured not by how fast they completed the test, but by how far they managed to get, i.e., the best would go all the way ana the middle ability ones would be stopped somewhere along the sequence. This modification in our own classroom testing patterns would be a big step forward in the direction of making calculus into a conceptual course, and there is no one stopping each instructor from doing this on his or her own. However, the whole department may want to take a unified move in this direction and prevent the students from searching out the "easiest" professor.

Microcomputer graphics should dramatically change our approach to graphs in calculus instruction. At this point in time, we use the calculus to get the information that is necessary to build the graph of the curve. After the picture has been sketched, the discussion terminates and no real global discussion of the nature of the process is given. It really should be the other way around. We can now start with the picture in a few seconds, and then time can be spent on the dynamics of the curve with the calculus information being added to enrich and amplify the points that are being made.

Much has been said about inexpensive calculators with definite integral keys and it should be clear that collectively we have to take a very critical look at the parts of calculus that have as their singular goal the finding of values for definite integrals.

Riemann Sums should be given a meaningful treatment with fresh examples and exercises. The use of averages and the expansion of the integral as an average concept looks like a productive path. Parts of the volumes of revolution section could easily be scrapped and replaced with a variety of applications of Riemann Sums from topics other than area and volume.

The current calculus course suffers from poor textbooks, disinterested students and misguided teachers. But calculus is still an intellectually strong and important discipline and the course survives in spite of the damages that we have made. If mathematics is indeed the subtle study of simple things, then should not we take this strong and simple course back to its subtle root .... the mathematical study of change.

REFERENCES

1. Committee on the Undergraduate Program in Mathematics, "Recommendations for a General

Mathematical Sciences Program," Mathematical Association of America, Washington, 1981.
2. Fey, J. T. (ed) "Computing and Mathematics," National Council of Teachers of Mathematics, Reston, VA, 1984.
3. von Braun, W., Private Speech, Huntsville, Alabama, 1964.
4. "Wall Street Week," Public Broadcasting System, Owens Mills, Maryland 21117.

ON THE TEACHING OF CALCULUS

by

Peter D. Lax

About fifteen years ago, when I first taught a calculus course with computing
and applications, the teaching of applied mathematics had a very precarious place
in the college curriculum. It seemed to me then, and still does now, that the
teaching of calculus is the natural vehicle for introducing applications, and that
applications give the proper shape to calculus; they show how, and to what end,
calculus is used. Without them a calculus course is in danger of resembling a
guided tour through a carpentry shop, with instruction on how to use each tool
(including some antiquated ones), but giving no sense of how to use them to build a
thing of beauty and utility; or a music class where most of the time is spent in
practicing scales and finger exercises, with little chance to listen to or play
a composition much above the level of chopsticks; or a language class where grammar
and syntax are taught systematically, but where there is little conversation,
composition or reading of literature.

Alfred North Whitehead thought that the retention of inert material in the
curriculum does the greatest harm to education. Calculus, as currently taught, is,
alas, full of inert material, which will remain there as long as the teaching of
calculus is controlled by the establishment, i.e. the group presently entrusted
with teaching it. There is a reason for an establishment; calculus is a very big
enterprise, taught to a very large number of students with diverse needs and
backgrounds; it is hard to resist the temptation to design a course that can be
taught with equal ease by a senescent member of the department, by an adjunct
picked up on the first of September, or a graduate student recently arrived from
the Orient.

The only way the situation can be remedied is to entrust the teaching of
calculus to those who actively use it in their own research. This is what

happened in the last 30 years to the introductory courses in physics and chemistry, but is yet to happen to calculus. To be sure, there has been some advance; some new topics, notably probability, have been added, and attempts have been made to infuse more rigor.

I turn now to a brief list of topics and attitudes that I feel are given undue prominence in the neotraditional course.

There is too much preoccupation with what might be called the magic in calculus. For instance, too much time is spent in pulling exact integrals like rabbits out of a hat, and, what is worse, in drilling students how to perform this parlor trick. Summing infinite series is another topic that has the aura of a magic trick, and is overemphasized at the expense of the concept of approximation and iteration of functions.

I feel that rigor at this level is misplaced; it appears as an arid game to most of those who comprehend it, and mumbo-jumbo to those that don't. Besides, many of the tricky proofs can be avoided by starting with sensible definitions, such as functions that are uniformly continuous over a closed interval, instead of functions that are merely continuous at every point of it. There is no need to prove in the first two years of calculus that the latter class is included in the former. Similarly, the fundamental lemma of calculus -f is constant if $f' \equiv 0$ - is easy to prove and intuitively clear for functions that are uniformly differentiable, i.e. for which the difference quotients tend uniformly to f'; there is no reason to prove this in a calculus class for functions assumed to be merely differentiable at every point. Even more extreme examples can be given involving functions of several variables.

I turn now to some examples that illustrate the role of computing in teaching the concepts and uses of calculus.

Many students have difficulty grasping the idea that the integral of a function over an interval is a number. The reason is that this number is difficult to produce by traditional methods, i.e. by antidifferentiation, and so the central idea is often lost. Numerical methods have the great virtue that they apply universally.

It was Newton's sublime discovery that the laws of nature (well, most of them) take the shape of differential equations. When special methods are introduced to deal with each one of the pitifully small class of equations that can be handled analytically, students are apt to lose sight of the general idea that every differential equation has a solution, and that the solution is uniquely determined by its initial data. Numerical methods are universal and give students an intuitive grasp of existence and uniqueness of solutions. Even more important is the ability to explore the qualitative and quantitative properties of solutions and discovering experimentally limit cycles, damping, fixed points, stability, loss of stability, etc.

Computer experimentation is an indispensable companion for the design and analysis of algorithms for numerical integration, for solving differential equations, for finding zeros, maxima, etc., Nothing convincees students (and nonstudents) of the superior efficiency of a clever algorithm over a routine one than seeing the one outperform the other in actual calculations.

How and to what extent should computing be part of a calculus class? What kind of computing facilities are needed? There are no universal answers and there will never be any; as the computing literacy of the population shifts - and that of the the students is likely to shift faster than that of the teachers - different alternatives will emerge. At any rate I do not wish to see the classroom transformed into a computer lab; blackboard and chalk, pencil and paper are and will remain essential tools of teaching and learning. And I certainly don't want to see the magic of canned programs and fancy graphics replace the other kind of magic decried earlier.

I like to start the first calculus class with a calculator in hand, and point out all the functions that the calculator is able to evaluate to eight significant figures.

I describe the Newton algorithms for taking square roots and cube roots, and demonstrate their efficacy on a few examples. Sooner or later one of the students asks where these clever algorithms come from; this answer is: from knowing calculus.

As a useful by-product of this demonstration, the students become familiar with the phenomenon of convergence, and are well prepared for the concept.

A hand-held calculur is useful in demonstrating the power of a highly accurate integration scheme when few points are used. But beyond that, when more elaborate calculations are needed to drive a point home, when tables and graphs become indispensable to digest the meaning of a calculation, we must rely on a minicomputer and adequate ways of displaying results.

Happily, reasonably powerful minis are now available, at reasonable cost, so that most students can gain access to them. One way to make them available is through a computer lab; an informative article by Breuer and Zwas on how to use such a lab has appeared in the September 1985 issue of **SIAM News.**

In the future - partly here already - we can look forward to a bewildering variety of software packages performing and displaying the operations studied in calculus. Some of this will be tied to a test, others not; some of it will be available for a fee, others gratis. All conceivable hardware and programming language will be enlisted. I hope this workshop will generate discussions of what is desired, and suggest some mechanism for disseminating information on what is available.

I would like to close by taking a couple of swipes at two recent educational proposals that bear on the teaching of calculus. They are made by two well-meaning groups of enthusiasts whose view of mathematics is woefully narrow. The first advocates teaching all concepts algorithimically. This seems to me wrong; for a concept, when presented properly, can be absorbed as a whole, while an algorithm remains a sequence of steps. It is only after a concept has been understood, and made part of one's thinking, that we turn to the intriguing task of devising efficient algorithms.

The second proposal advocates diminishing the emphasis on teaching calculus in favor of teaching discrete mathematics. No doubt, discrete mathematics is a collection of beautiful subjects, many of which have gained great importance because they are intimately related to computers; but it is rubbish to say that calculus based mathematics has become less important. I have described elsewhere the recent explosion of theoretical developments in calculus based mathematics; here I want to point out that at the birth of calculus Newton showed how to write down equations of motion for systems governed by the most complicated concatenation of forces. But he did not show how to solve these equations, except for a small but significant class of problems. That today we can use computers to explore the behavior of solutions of all such equations is truly revolutionary; we are only beginning to glimpse the consequences.

Reflections of an Ex Foundation Officer

Stephen B Maurer
Department of Mathematics
Swarthmore College
Swarthmore PA 19081

During 1982-84, I had the privilege of serving as a Program Officer at the Alfred P Sloan Foundation in New York. It was a privilege because, in attempting to educate myself to make informed judgments about funding requests, I had a much better opportunity than most mathematicians to develop a national perspective on problems and opportunities in the mathematics profession.

My goal in this paper is to try to share some of my perspective, as it relates to the teaching of calculus. To that end, the organization of this paper is as follows. After some introductory remarks I will turn to a listing of the key aspects of the current environment, as I see it. Then I will discuss some of the responses I hope this conference will make.

I wish I had a clearer perspective than I do -- It would be nice to have one theme to drive home. Unfortunately, things aren't so simple, and this paper is rather discursive.

To begin with, many of you may be wondering what I am doing at this conference. I am best known as an advocate of discrete mathematics, to the extent of urging that it replace a certain amount of calculus. So, did Ron Douglas invite me as a saboteur?

Of course not. I have been advocating discrete math because I think it is pretty and important mathematics, and because the weight of tradition will relegate it to the periphery unless someone advocates it. But the fact is, by my own wish much of my teaching has always been in continuous mathematics, especially calculus.

I am at this conference because the same general developments which lie behind the discussions of discrete mathematics are crucial for fresh discussions of calculus. If there are changes to be made in calculus instruction, and I believe there are, the same weight of tradition will bury them unless many of us speak up.

The Sloan Foundation too, because of its recent efforts in discrete mathematics, may be thought by many to be an unlikely source of funding for this conference. Not so. Although I am no longer their employee, I think I can say with confidence that their interest is, and always has been, broad. Their goal is to encourage the best of ideas in mathematics and mathematics education -- to the extent that they can with a small part of a small budget (relative, at least, to government funds). In particular, having funded a part of the Reformation, Sloan is happy to consider funding part of the Counterreformation as well!

One more introductory idea. I crib this from another article of mine [7], which deals specifically with the prospects for more discrete mathematics. I regard that article as a companion to this one, because many of the same general background issues are discussed. In any event, one of the things I did in preparing that article was to review a classic hard calculus book of the early 60's: Apostol [4]. That review led to a new insight: "It's not a calculus book. It has all sorts of things in it that we discrete math advocates have been talking about -- combinatorics, induction, probability and statistics, difference equations, numerical analysis. Granted, much of this is in the second volume, certain current perspectives and topics are completely lacking (the algorithmic perspective, the topic of graph theory), and it is all wrapped around calculus as the core. Nonetheless, it's clear that Apostol wasn't trying to teach just calculus; he was trying to teach mathematics, as an integrated two-year course, according to the best judgment he could make at the time."

I think our present conference is based on the premise that there still can be an integrated two-year course in mathematics, appropriate for the majority of future users of mathematics, and that calculus can still be the core. At any rate it should be the premise of this conference, even if I personally might debate the unique centrality of calculus. The issue for the conference then becomes: what renewal of the calculus is necessary to make that core as vibrant as it should be?

What's Going on out There

In light of the previous paragraph, the most important observation I can report is that

1) The idea that there can be an integrated, general purpose program in mathematics for the first two undergraduate years is under attack.

The attack is not a conscious effort based on disagreement in principal with the goal of a common sequence. Rather it is the result of various structural aspects of undergraduate education.

First, as more and more students take mathematics, there is a natural tendency to create different courses by interest groups -- engineering calculus, social science calculus, math for business majors, math for computer science majors. Second, lack of interest by mathematics departments over the years in servicing groups who weren't considered talented or to have "real interest" in mathematics itself has led to other departments teaching their own math courses for their students. Such offerings come to have less and less in common. Finally, the general view in computer science departments that discrete math is more important than calculus (with some departments not

requiring any calculus at all) is perhaps a larger threat to a
common curriculum than everything else heretofore -- because the
computer science crowd includes a lot of smart students who would
otherwise take the mainstream math courses, indeed, might
otherwise become math majors. Furthermore, there are lots of CS
students, and the CS departments, unlike the psychology
departments, don't want to teach their own math courses -- they
don't have the staff. So the math department has a service
course which begins to overwhelm the previous core courses in
enrollment.

Admittedly, some people don't feel that the growing
fragmentation of offerings is much of a problem. Is not
diversification a natural and good concomitant of growth and
evolution? But I do very much feel this is a problem.

Partly I feel this way because of my background -- I was
educated at a small liberal arts college and teach at one. We
don't have the enrollments or the staff to design a large variety
of introductory courses. More important, we are against it in
principle. The Liberal Arts creed is that one is best educated
for a useful career by being educated for no career in
particular. And this works -- so long as the career-independent
education in fact involves a lot of learning which is broadly
useful in many careers. If the mathematics which gets taught
ceases to be broadly useful, then liberal arts mathematics
education is in trouble.

Finally, there is a major problem with diversification of
offerings for the students. They have to choose their futures
earlier and earlier. For instance, if CS requires a year of
discrete math by the end of the sophomore year, and mathematics
requires instead linear algebra and multivariate calculus, then
most students must commit themselves one way or the other by the
beginning of their sophomore year. If CS requires discrete math
in the freshman year, and the math department requires calculus,
the dilemma is worse.

The question of core curricula comes up in other disciplines
as well. Indeed, a few years ago, my college considered
introducing a required core curriculum for all students. Despite
what I have said above, I was against this. Indeed, I argued
that trying to show students unity in overall intellectual
endeavors was precisely the wrong thing to do. What people need
to learn is how to cope with increasing disunity, both
intellectual and social.

Yet in mathematics I do think some standardization in our
introductory offerings is much needed. This is because
mathematics is a basic skills area. If you don't learn any
probability or statistics, and aren't among the few bright enough
or eager enough to pick it up later on your own, you really will
be at a loss for various purposes. If you learn only discrete
math and not continuous (or vice versa), you really will be

cutting certain options out. Whereas for history, you can take
your first course at the end of your sophomore year and still
major in it, and, if that first course is ancient Chinese history
you can still go into a modern European upper level course
afterwards, in mathematics such things don't happen.
Furthermore, mathematics is not a subject students pick up again
later in their studies. On the contrary, the problem is that too
many stop after one course, too many more after two courses, and
so on. For all these reasons, some sort of agreed upon central
curriculum in the first two years is very much to be desired.

So much for a sermon, one which I hope was actually
unnecessary for this audience. But keep in mind: if we want a
common curriculum that most mathematically able students will opt
to take, it's got to touch more bases than the traditional
curriculum.

2) Mathematicians say that what they are really
   teaching is how to think, but actually, we are more
   technique and fact oriented than many other
   disciplines.

While at Sloan I didn't work just on mathematics issues. I
saw proposals from many disciplines. One of the things I learned
is that the self-justification "we teach students how to think"
is not the private preserve of mathematicians. Almost every
discipline asserts it. But if it's modes of thought rather than
specifics that a discipline considers important, and all their
courses teach it, then almost any course ought to be an entrance
into the major. As remarked above, this is true of history. It
is not true of math.

To cite another example, if you ask a mathematician what he
accomplishes in some course, he will usually answer by telling
you the theorems he gets to, not the methods of reasoning he
tries to instill or the competencies he tries to impart.

I'm not saying we should make all courses equal as far as
entering our departments is concerned, or that we should stop
teaching key theorems. But I am saying that our image of
ourselves as educators is erroneous.

3) There are several "standard" versions of the
   calculus sequence today, running from 2 to 4
   semesters.

Historically there were 4 semesters. Then in most places it
moved to 3 to make room for linear algebra. In some places, it
is now two, by which I mean one gets up through the basic
concepts of multivariate calculus in 2 semesters. Some of the
2-semester places are MIT, Carnegie Mellon, Dartmouth, Williams,
Haverford and Grinnell. Sometimes this is accomplished by going

fast and having 4 or 5 meetings a week, and/or having very bright students). Sometimes it is accomplished by "kicking upstairs" some traditional material. For instance, at Grinnell <u>all</u> material about sequences, series and power series is deleted from the calculus course. It reappears, combined with material on numerical analysis and differential equations, in a sophomore-junior course taken mostly by math and physical science majors.

There is even more diversity in the relation of discrete to continuous. At many places discrete math is first offered at the junior level. On the other hand, at a number of schools it is now required before calculus. (My impression is that the latter schools have weaker students, mostly interested in computing, so that discrete math serves in part as a fresher version of pre-calculus and in part as pre computer science.) Finally, there are a few places where some calculus and some discrete math are joint prerequisites for further mathematics of any sort. For instance, Dartmouth now has a system where a semester of calculus and a semester of discrete are required for all further work in math or computer science.

In short, there are many experiments going on with calculus and with the mix of subjects in the first two years. Nonetheless, a traditional calculus sequence, with a traditional curriculum, is well entrenched. It is not hard to understand why. Calculus is the least denominator. Since everyone knows it, you can have anyone teach it -- your least innovative or least energetic department members, even your graduate students who don't speak English. But, if everybody is going to teach it, indeed teach it simultaneously, then you've got to stick to common ground. Ergo: nothing changes.

4) Most calculus books have changed very little.

This least denominator argument above applies equally well to books. Or maybe I should say greatest denominator. For a book to sell well, it must include not only whatever <u>every</u>one regards as traditional, but also whatever <u>any</u>one regards as traditional. Indeed, if typical books have changed in the last 10 years, it is by the addition of new business, social science and biological science applications, and more material on numerical methods, without the elimination of any physical science and engineering applications. All this makes for giant books, and thus makes it harder for faculty to decide that large chunks of material don't really need to be taught anymore.

This is not the whole story. There is a whole class of calculus books that hardly existed 20 years ago -- the exclusively business/social science/biological science oriented books. In many ways these are interesting books. First, they are much shorter -- they show that it <u>is</u> possible to delete a lot of traditional material. Second, a few of them are quite well

written. (I happen to like [5], but this opinion is based solely
on browsing, not teaching.) The good books succeed in motivating
the material and showing that deletions do not necessarily mean a
course without conceptual content. Third, a few of them are
nicely experimental. For instance, [11] mixes a certain amount
of discrete math into his calculus by including a recursive
approach to sequences, a sequence approach to limits, and a
pairing of difference and differential equations. (I have a
report, however, that this is a horrible book to teach from.)

There is second new class of calculus books -- those for
weaker students. Unfortunately, these two new classes are more
or less the same. This is unfortunate for two reasons. First,
this tends to perpetuate the division into careers by ability.
Second, it perpetuates in the minds of faculty the idea that only
the thick books are the real calculus.

I do know of one thin calculus book that cannot be faulted
for being wishy washy, [6]. So perhaps this is an existence
proof that a short, solid calculus is possible. But I'm not
happy with that book either -- in addition to throwing out all
the special topics, it throws out the motivation and any sense
that the subject is applicable.

Finally, in the 70's there was a spate of books with titles
like Calculus and the Computer [9]. These arose from attempts to
teach calculus with a heavy computing orientation. Most of these
books have died out. Perhaps they came too soon; in those days
one needed a mainframe. But perhaps they failed because faculty
felt they took too much attention away from the core issues. I
don't know. But this does lead me to the next item.

5) Now computers can do symbolic manipulations as well
   as numerical ones, and even microcomputers can do
   the former moderately well. Furthermore, the
   mathematics community now seems pretty well aware of
   this development, although only a fraction of
   faculty have actually tried out the software, and
   very few courses use it.

I'm less prone now than two years ago to make visionary
statements about the great change such "computer algebra" will
cause. First, it is not likely to have much impact on classes
until it is available on hand calculators. (On the other hand,
it is already having considerable impact on the daily life of
engineers and on the research of many academics.) Second, it is
unlikely to save much time in courses. Perhaps, as with today's
hand calculators, after an initial period of fuss computer
algebra on calculators will have little effect on the "daily
life" of calculus classes. However, I hope not. I really do
hope such software will allow us to accelerate a change towards
more emphasis on ideas and less on techniques. But this is a
hope. As far as what's going on out there, the answer is: the

jury is still out -- in fact, it just went out.

6) Academic mathematicians are not very happy these
   days.

Salaries are down in real dollars from before the
inflationary surge. Class sizes are up in elementary courses,
while upper level courses often languish. The students are often
abysmal. The graduate students who often teach the elementary
courses aren't much better and there aren't many of them. The
amount of money available from one's school for travel and
professional activities, let alone for phoning, xerox and office
supplies, is very tight. Finally, grant money for research is
hard to get, and federal money for curricular development at the
undergraduate level is nonexistent. It does not seem that the
professoriate gets much respect.

This is not particularly a problem of mathematics -- it is
academia wide. However, it is the general perception among
academic mathematicians that they are suffering more. There may
be truth to this. The surge in introductory enrollments is not
universal. At the research level, the David Report [3] has made
the case that mathematics is getting a worse deal. Perhaps a
case can be made at the undergraduate level as well. But it's
got to be made well, because Deans, college Presidents and
funding agencies are receiving special pleas from every corner.
Their usual reply, quite understandably, is that everybody's got
to share the dearth.

Are our special pleas legitimate? An effort is underway to
find out. As a follow up to the joint AMS/MAA/SIAM Committee on
the Status of the Profession, there is now an MAA Planning
Committee for a National Study of Resources for Collegiate
Mathematics. Recently (September 1985) the MAA has received a
grant from the Sloan Foundation for this Planning Committee to
work up a proposal for a major study to be made by the National
Research Council. If such an NRC study comes about, it may well
be the equivalent of a David Report for undergraduate
mathematics, with similar results -- more attention and more
money for undergraduate mathematics. Concurrently, the National
Science Board (the board of trustees of NSF) is reviewing the
wisdom of the current allocation of federal math funds into
research and precollegiate education only.

7) The use of mathematics in other disciplines is
   changing greatly. Mathematicians don't have a clear
   enough idea how.

For instance, in my own institution, introductory physics is
very different from when I took it. Simple examples are still
done in closed form, but after that everything is done with
(discrete) computer approximations and graphics. Similarly,

engineering students at Swarthmore solve linear systems, both algebraic and differential, with computers in their first engineering course.

I'm sure that, at the research level, a physicist or an engineer can still use excellent competence at calculus techniques -- and much more advanced analysis. (I am told, though, that even a lot of theoretical physics is now done on discrete models, models which sometimes do not have continuous models as their limiting case). However, if the needs for in-depth calculus are at an advanced level, then the in-depth treatment need not be included in the first calculus course as far as physics students are concerned.

From the Williamstown conference on discrete math [8], at which representatives of other fields were present, I also know that at least some people in all the traditional client disciplines feel that a very different mathematics preparation would be fine.

In short, I know just enough to know that I am eager to hear what the representatives of other disciplines at this conference have to say.

Once we hear what they have to say, we still have the problem that not all applications can be included. Quoting again from [7], "There is a whole tradition of expectations about what students will learn in their first two years of mathematics. .. a physics professor can assume that a student with 3 semesters of calculus knows Green's Theorem, but an economics professor cannot assume the student has even heard of Euler's Theorem on homogenous functions."

These traditions no doubt made sense at one time -- the applications chosen were the ones for which there was the most demand. But the relative demand may have changed without our knowing it. And with many more disciplines using mathematics, relative demand may be harder to assess.

Let me suggest a rule for deciding what to include. It will sound innocuous enough, but it may be impracticable. If an item (theorem, technique, problem type, etc) is of considerable interest to just one discipline, exclude it from this first course and tell that discipline to teach the item itself; otherwise consider including it.

What This Conference Should do

Thinking up my answer to this question has been useful in getting me to focus my ideas. Perhaps a similar exercise will be useful for you. At any rate, if I could wave a magic wand and get this conference to do what I wanted, here is what it would be.

1) The conference should call for further study of whether mathematics teaching suffers more than other disciplines, but it should not waste time reciting our current woes.

We've all heard our complaints about calculus, and so have the powers that be. As a conference, I don't think we will be in any position to provide the hard data needed to determine if our plight is different than other academic plights. At most we can help point out what data is needed. Thus, we can add momentum to the efforts toward such a study, but that's it.

In some sense, calculus is the subject mathematicians love to hate. If it didn't exist, we would have to invent it, as an outlet for our need to kvetch. But while this is a great coffee room avocation, it doesn't make for a good conference.

2) For the short run, this conference should call for projects which require only small amounts of money.

It is necessary to think in terms of help we can reasonable expect to get. To say that class size ought to be cut in half is useless. No foundation in the world has the funds to underwrite that, even if the additional faculty could be found. (On the other hand, if a case can be made for a unique distress in mathematics, in the long run academic budgets might be moved to improve the classroom ratio.) Similarly, it's no use telling publishers to change their ways. They are hemmed in by market forces. We must show them that a new type of book will attract a market before we can expect them to help.

The amounts of money likely to be forthcoming from private foundations are small, say, in the tens of thousands of dollars. What can be bought with that sort of money? The answer that occurs to me is faculty time -- to try new versions of a course or new classroom techniques, to write experimental materials, to compile another report. We should try to think of other answers and make recommendations concerning the answers we like best.

3) Don't start from scratch in our recommendations but build on work already done.

Coffee room complainers have the luxury of ignoring or belittling what has been thought in the past. We do not. There is a long history of thought about calculus. For instance, are large classes really hurting calculus? It's easy to simply assume so, but in fact there have been a number of studies and the conclusions are mixed. The studies may in fact be flawed, but if so we have to indicate why and suggest what further study should be done.

Similarly, if we make recommendations for the calculus

syllabus (as I hope we will), it won't be as if this has never been done before. So far as I know, the most recent national effort on the calculus curriculum is in the Tucker report on the Mathematical Sciences [1]. Years earlier there was a CUPM report on calculus [2]. We should start, I think, with the Tucker report and see how we want to change or amplify things.

4) In terms of the syllabus, think hard about what can be taken out.

It's easy to talk about what should be covered in the calculus and isn't. (I plead guilty to doing a bit of this myself elsewhere in this paper). But this approach has led to our current large books. It's much harder to bite the bullet and throw things out. But that's what needs to be done -- in great measure if the goal is to shorten the course, in considerable measure if the goal is simply to modernize it. To talk instead of "deemphasizing" this or that doesn't do the trick: as long as some topic has a toe in the door, it's impossible to make sense of it for the students without devoting a certain amount of time. Usiskin has written a very interesting article [10] on what not to teach in high school algebra and geometry. We need a similar effort for calculus.

5) Think more in terms of competencies than in terms of theorems.

This harks back to my remark that mathematicians talk about teaching to think but describe their teaching in terms of topics and theorems. This tendency can be counteracted in part by putting some emphasis on the competencies. For instance, I would like to see

-- learning how to write coherent problem solutions, including correct mathematical sentences

as one of the competencies for a first calculus course. I would like to add

-- learning how to model a problem with calculus, and

-- how to apply algorithms to obtain approximate answers.

I would like to downplay "how to compute the answer in closed form".

6) Seek a clear idea of what other disciplines need from us.

I'm looking forward to what the representatives of other disciplines at the conference will have to say. But I also know we must probe deeply. Generally, people in other disciplines haven't had any reason to think deeply about how they really use mathematics in their teaching, and perhaps even in their research. Also, the representatives at our conference are necessarily few in number. What we glean from them must be viewed as part of a larger effort to keep in touch with other disciplines on educational matters -- an effort in which the MAA has ongoing activities.

   7) Write a syllabus (or several) but make them for a
      course in mathematics, not just calculus.

This harks back to Apostol's goal of teaching mathematics, and to my hope for a standard, integrated curriculum. To begin with, I hope the conference will decide whether it approves of the common curriculum idea or not. If it does, it should try to make its calculus course serve a more general purpose. For instance, probability is very important; some uses of calculus in probability ought to be introduced, even if the probability must be treated informally. And it shouldn't wait till the second volume. Similarly, given the increased interest in discrete mathematics, opportunities to mix the continuous and the discrete ought to be seized when they can. Thus there ought to be something on how power series are also generating functions, and partial fractions might be introduced there instead of (or in addition to) in methods of integration.

I've reached the end of what I wanted to say. I wish I had a grand conclusion for you, but my paper just trails off. In doing do, perhaps it will remind us that the discussion of calculus is an endeavor which will never have an end.

References

1. "Recommendations for a General Mathematical Sciences Program," (The Tucker Report), Committee on the Undergraduate Program in Mathematics, Mathematical Association of America, 1981.

2. "A Compendium of CUPM Recommendations," Vols 1 and 2, Committee on the Undergraduate Program in Mathematics, Mathematical Association of America, 1979.

3. "Renewing U.S. Mathematics: Critical Resource for the Future," (The David Report), National Research Council, Washington, 1984.

4. Tom M Apostol, "Calculus," Vol 1, 2nd ed, 1967 and Vol 2, 2nd ed, 1969, Wiley, New York.

5. Larry J Goldstein, David C Lay and David I schneider,

"Calculus and Its Applications," 3rd ed, Prentice-Hall, Englewood Cliffs NJ, 1984.

6. Serge Lang, "A First Course in Calculus," 2nd ed, Addison-Wesley, Reading Mass, 1968 (316 pages).

7. Stephen B Maurer, The lessons of Williamstown, in "New Directions in Two-Year College Mathematics," Donald J Albers, Stephen B Rodi and Ann E Watkins, eds, Springer-Verlag, New York, 1985. pp 255-270.

8. Anthony Ralston and Gail S Young, eds, "The Future of College Mathematics: Proceedings of a Conference/Workshop on the First Two Years of College Mathematics," Springer-Verlag, New York, 1983.

9. David A Smith, "Interface: Calculus and the Computer," Houghton Mifflin, Boston, 1976.

10. Zalman Usiskin, What should not be in the algebra and geometry curricula of average college-bound students?, Math Teacher 73 (Sept 1980) 413-24.

11. Thomas Wonnacott, "Calculus, an Applied Approach," Wiley, New York, 1977.

STYLE VERSUS CONTENT: FORCES SHAPING

THE EVOLUTION OF TEXTBOOKS

by

Peter Renz

SUMMARY

To focus on matters of content to the exclusion of matters of
style and execution in proposing new curricula is to lose sight
of the essentially evolutionary nature of the development of
courses and course materials. Strong selective pressures work
against large-scale changes in large courses. These constraints
are partly the result of inertia and of existing requirements
outside of the mathematics curriculum, and they are partly
economic in nature, affecting what will and what will not be
economic to publish. This is a review of these limitations on
curricular reform and the conclusion is that the success of
such changes is more dependent on the style and execution of
the materials produced than on the exact content of those
materials.

85

STYLE VERSUS CONTENT:
Forces shaping the evolution of textbooks

Peter L. Renz
Division of Science, Bard College
and W.H. Freeman and Company

STYLE AND CONTENT

What is it that makes a truly successful course or text?
Is it more a matter of content or of the manner in which
the material is presented? The answer depends upon one's
objectives and one's point of view. On balance I say that
style and manner of presentation are more important than
content. In particular, the matter of discrete versus continuous mathematics seems a side issue. The main issue should
be how to achieve success in the classroom, and this issue
depends upon the details of the teacher's interaction with
his or her student. Where texts and courseware are concerned,
success depends upon the ability of authors to reach out to
students with apt and compelling arguments and with clear and
evocative images. Success in education also depends upon
students and teachers having clearly formulated goals, both
overall and in detail, and ways to judge the level of success
in achieving these goals.

One might properly be alarmed or horrified by the content, or
even the methods, of R.L. Moore's topology classes. One
might even disapprove of the narrow focus that comes with the
use of pure discovery-learning, but Moore's successes and
those of his students are legendary. This, therefore, shows
that style can triumph over content. I venture that there
are no examples of the triumph of content over style. If
the presentation is sufficiently bad, the students are lost.

The debate concerning whether discrete or continuous mathematics should be central to the curriculum will simply pass away. Change will come by evolutionary forces. Those who heralded the discrete revolution will be enshrined as saints by some and cursed as devils by others. But the changes that are made will be the work of the foot soldiers, teachers in the trenches who are subject to unpleasantness and risk as they work up new course material or help purge the errors and infelicities of other's course notes. These teachers are my heroes, along with their striving and suffering students.

The discrete revolution is being oversold. I raise three objections to it as a cure-all. First, the discrete/continuous dichotomy is not as sharp as it is pictured. This is borne out by many sources, but I have been particularly impressed by the depth and breadth of the arguments made in response to Anthony Ralston's position piece to appear in the November 1984 issue of The College Mathematics Journal. The responses by James P. Crawford, Daniel J. Kleitman, Peter D. Lax, Saunders MacLane, Daniel H. Wagner, and R.L. Woodriff emphasize the central importance of the insights gained from continuous mathematics. Insights of great use even for the study of discrete systems. Such respondents as William F. Lucas and R.W. Hamming stressed the importance for modern applications of both discrete and continuous mathematics--and these respondents are very strong proponents of discrete mathematics.

Second, the importance of manner of presentation of the material has been largely pushed aside in the struggle over what is to be presented. My thoughts on this have been sharpened by seeing how clearly this issue has been set forth by the responses to Fred Roberts's position piece in the cited issue of The College Mathematics Journal on the role of discrete mathematics in the college curriculum. Of the six responses that I have seen in draft, four see the main issues in the introductory college mathematics curriculum as being pedogogical (style and manner of presentation) rather than content. I direct your attention to the responses by John Mason, Patrick W. Thompson, and William Ellis, Jr.

Third, until strong and successful models for these new discrete mathematics courses are available, we do not even know
exactly what is being proposed. The reason is that discrete
mathematics is a very broad area and it is full of very difficult and demanding material (try to master Ramsey theory, for
example). Until a practical course has been plotted through
these seas, the proposal that this voyage be undertaken by
large numbers of teachers is as irresponsible as it would have
been to propose a general assault on the problem of sailing
west from Europe to India before Columbus, Magellan, and others
had led the way.

For these reasons and for others set out below, I believe that
we should put the noise of debate behind us and get on to the
real business at hand: making the experiments that will give
us new courses for this new curriculum and make safe the
voyage to this new land.

My plea is for evolution not revolution for two reasons: first,
evolution is the way things actually happen and second, evolution is a continuing process. Change is essential if mathematics is to be a vital subject.

DEVELOPING NEW COURSES

The development of new courses from their conception is similar
in some respects to the emergence of a new species. Ideas serve
as a modifiable genetic code guiding the development of the
course, but these ideas are not enough. There must be the
proper local environment to allow realization of the idea;
and, if anything is to come of this all, the idea and its
realization must be able to catch on elsewhere. The proposal
to restructure the college curriculum so that discrete mathematics is taught early and calculus comes later is like proposing to insert a new gene into the chromosomes controlling the
cells of the curricular organism. I believe that a cautious
approach should be taken to such experiments lest one produce
monstrosities.

Natural mutations yield many variant forms, but those that

represent substantial variations of large creatures naturally
abort and never live to see the light of day. Successful
variants begin small. Mammals first appeared as a few small
creatures and through evolution, with its general tendency
toward larger forms (Cope's rule) and greater diversity and
specialization, gave us the full range of mammalian life that
we see today. For creatures the size of elephants, we see
little in the way of rapid change. (See McMahon and Bonner
(1983) for the biology.) So it should be with curricular
change. The "new mathematics" experiment illustrates the disastrous potential of large scale, rapid, and radical change
imposed from without.

The wonderous diversity of life arises from selective pressures
acting on random mutations. We may best understand the
dominant types of texts and courses as rising from a similar
evolution under competitive pressures. This viewpoint explains
much, even though it ignores the distinction between the
wholly random character of mutation in biology (prior to the
invention of genetic engineering) and the purposive nature
of curricular reform, and course and textbook planning.

As an editor, it has been my task to anticipate curricular
developments and to select or develop texts that will establish
substantial positions in both existing and new markets. From
the editor's point of view, it is clear that there are more
sound, even exciting possibilities than there can be economic
niches in the market. These possibilities, whether put forth
in book proposals, class notes, comprehensive plans for curricular reform, or in discussions, are just so many hopeful
mutations whose potential is yet to be proved. This presumes
that we know what success means in this context. For publishers, success is a commercial matter—sales great enough to cover
costs and give a pleasing return that can be used to finance
new projects, provide increased compensation for employees
and increased returns for investors, and meet all the other
needs of a prosperous enterprise. What publishers seek is
overall success-that is, success of the enterprise as a whole.
This is by no means inconsistent with taking on high-risk
projects of occasional projects primarily for the good of the

discipline and as a sign of the house's commitment in a given area.

Academics and publishers often do not understand each other simply because they do not understand each other's goals or their separate conditions of life, including the constraints on achieving their separate goals. Here I will give the picture as seen by a publisher. I will focus on the numbers, because they are easy to understand and they are the main constraints affecting all publishers almost equally. Moreover, the numbers will show where the best opportunities for curricular reform lie, insofar as realizing those opportunities depend on the publication of new course materials

WHY LARGE TEXTS FOR LARGE MARKETS SHOW LITTLE VARIABILITY: THE EFFECTS OF SCALE

Large technical texts with many diagrams are expensive projects even before the first copy is printed. Typesetting, art, page make-up, including allowances for all proof and corrections for a two color, 8 X 10 inch, standard calculus book of about 1200 pages would cost about $228,000 in 1984. To these costs we add editorial expenses including reviewing, checking of answers and solutions, travel., grants to the author for manuscript preparation, and in-house editing and development. I estimate these at about $50,000. Finally, for a complete package one must add the cost of supplements (Solutions Manual, Student Study Guide, and so on), a small publishing program itself. These might run to $40,000.

How does such a book look as a business proposition? The publisher must face the fact that competition, including the new and used book market, takes a very heavy toll on sales in the larger markets (annual decreases in sales usually exceed 20% and may exceed 30%). Furthermore, substantial sales for mainline books can seldom be sustained much beyond the fifth year after publication. Thus, a book that sells 10,000 copies in this sort of market in its best year might be expected to sell 30,000 to 40,000 copies altogether before it fades completely. Such sales would prove disastrous for the publisher. Here are the numbers for sales of 35,000 copies over 5 years at a nominal list price of $40.

Profit (loss) on sale of 35,000 copies of a calculus book
List price ....$40.00, Net Price........   32.00
Net receipts $1,120,000
Less printing, paper, and binding
costs at $5.11 to $6.50 per book............... $178,850
Less royalties at 16.5% of net $184,800
Less operating overhead $560,000
Less investment $318,000
                 Net loss...................... $121,650

Here I have allowed 50% of net receipts for all operating
costs under the assumption that all miscellaneous costs,
including those representing the time value of money are ex-
pensed for the publishing house as a whole. This allowance
is perhaps high, but it is close to actual costs, and it is not
clear that this allowance could be reduced below 40% for the
operations of any sizable publishing company. I have assumed
that inflationary and interest effects generally cancel each
other. The exact cost of the printing paper and binding depends
on the lengths of the printing runs. With a run of 30,000
copies the cost would be about $5.11 per copy; with a run of
10,000 or under copy cost could rise above $6.50.

For the publishing house to earn a profit of 10% of net before
taxes (a modest rate) with this model, it needs to sell about
60,000 copies. Again allowing 5 years for this would require
selling 17,000 to 20,000 copies in the best year. This represents very substantial success. In the real world few books
sell substantially above 20,000 copies in their best years.
The publishing house does not move into a really comfortable
position until sales of 30,000 to 50,000 copies are achieved
in a calculus text's best years. However, success in the
30,000 to 50,000 copies per year range allows important economies of scale. In particular, it floats a large operation
(promotion, sales, etc.) and it positions a company well for
revisions, which are generally less expensive (art can be re-
used, etc.) and more certain in pay-off.

Nevertheless, this analysis shows why in freshman calculus
(500,000+ students per year) and other large markets competition
produces uniformity. No publisher can afford to produce a
full-scale calculus book that is so different or so specialized
that it is not a serious contender for mainline courses.

In recent times, no slim alternative book has swept the market
or even taken as much as 5% of the market. But the costs of
launching a full scale calculus book can't be comfortably borne
by sales of less than 5% to 10% of the market. While the
upper end of this range, 10%, represents about the largest
slice of the pie that one might look for today. The result
is that calculus books tend to compete mostly through minor
variations and improvements. The dominant books expand by
spawning minor variants that act to cut down the profitability
of the market and thus squeeze out competitors. Consider
Swokowski's Alternate Edition.

Here we see Hotelling's law of duopoly at work. Large competitors compete by matching each other's products and going
for the center of the market. Consider Ford and General Motors,
Time and Newsweek. This is convergent evolution at work in an
area where the economies of scale allow only a few variants to
survive. The freshman calculus text is an elephant in the
world of texts and the economy simply can't support very many
types of elephants. Economic forces prevent substantial innovations in courses of this size.

CURRICULAR FORCES LEADING TO UNIFORMITY IN LARGE COURSES

Large courses are taught by teams and the direction of such
courses is determined by group choices. Committees work by
averaging and tend to conservative decisions. Moreover, large
courses in the sciences and mathematics usually have a strong
service function, both for higher-level courses in the same
discipline and for courses in other disciplines. The client
departments for such students expect a uniform and dependable
product, and these expectations strongly limit the possibilities
for innovation or change.

Redirection of the introductory curriculum requires agreement

on new standards. Until such standards prove themselves
workable in the classroom and are found more effective for
student's future work and for the client departments, prudent
departments will not rush to institute new courses and prudent
publishers will not rush to publish large or lavish books for
such courses. Indeed, even mild deviations that are untested
may be viewed with suspicion.

COMMERCIALLY ATTRACTIVE INNOVATIONS

Today's calculus book evolved from smaller ancestors by gradual
additions and adaptations. If the present type did not exist
for us to imitate, no author could or would invent it as an
entirity and no publisher would invest in it. However, new
types of texts are produced every year. Many are eliminated
by market forces. The quite considerable costs and development and publication are borne by the authors, their institutions, and the publisher. As an editor, I have seen that many
serious efforts are never pushed to completion by the authors
while others are found by reviewers to be unsuitable for their
intended markets and do not see the light of day. Each of
the following examples was innovative and commercially attractive. They run from small projects to immense ones, and they
illustrate what qualities innovative projects must have if
they are to be commercially acceptable.

Smaller projects allow more attractive opportunities for innovative publishing. For example, Loren Larson's ALGEBRA AND
TRIGONOMETRY REFRESHER FOR CALCULUS STUDENTS was based on a
successful supplement in use at the author's school. Neither
the idea of a review book nor the concepts and skills to be
reviewed (content) were novel, but the way in which the review
was set out was compelling and the care in bringing it together
let one see that this particular approach would work (style).
This project was attractive in its existing form to reviewers
at other schools. Furthermore, it made sense to print this
informal book from author-supplied copy and typed in final
form after the publisher's work on the art, design, and edit-
ings. The result was a project requiring an investment equal
to about 0.05 times that of the standard calculus book described
earlier.

To break even with this sort of small project, one need only
sell 3,000 copies. The sales of this book have exceeded
30,000 copies in five years and it has achived a percentage
return on net beyond that which could ever be expected in a
standard calculus book.

Other examples of similar success of small projects based on
successful texts in use at a local school, each of which meet
a need not met by existing texts, are QUICK CALCULUS by Daniel
Kleppner and Norman Ramsey (John Wiley and Sons) and OPERATIONS
RESEARCH FOR IMMEDIATE APPLICATION: A QUICK AND DIRTY GUIDE
by Robert E. Woolsey and Huntington S. Swanson (Harper and Row).
I would argue that the success of these books is more a function of their style and execution than of their content.

An innovative project of larger scale is David Moore's STATIS-
TICS: CONCEPTS AND CONTROVERSIES (W. H. Freeman and Company),
again it was based on a successful course at the author's
school and had proven attractive in manuscript form to those
experimenting with new types of statistics courses elsewhere.
This project represents an investment equal to about 0.10
times that of our standard calculus book and requires sales
of roughly 10,000 copies to break even. To achieve a 10%
return on net with this sort of project sales of 20,000 copies
are needed (roughly 0.33 times those required for our standard
calculus text). The slow pay-off here is a result of holding
the price down to allow supplemental use, a strategy that
succeeded very well. Larger still and equally successful in
its niche is the innovative statistics book by David Freedman,
Robert Pisani, and Roger Purves (W. W. Norton). This text
was developed and class tested over many years at the University
of California at Berkeley. It is style and execution not content alone that has made these books successful.

More pertinent to the discrete mathematics debate is MATHEMA-
TICAL STRUCTURES FOR COMPUTER SCIENCE by Judith Gersting (W.H.
Freeman and Company). Again, this is a text that arose from
an existing successful course at the author's school. Here
the target was the ACM's proposed course on discrete mathematics. The books cited earlier in this section were innovative

in their goals; this is not the case for Gersting. This course
existed elsewhere, enrollments were rising, and the author
was unsatisfied with the available texts. It was not a matter
of what the available texts set out to do (content) but rather
how they did it (style). Gersting's course notes were polished
during more than three years of class testing before the book
was published. Her class notes were extensively reviewed by
other teachers to ensure that the local success could be
repeated at other schools. The result is a text that requires
an investment not much above 0.10 times that of our standard
calculus book and that shows a profit of more than 10% of net
receipts with sales of less than 10,000 copies. That is about
0.16 times the sales required for equal profit from a calculus
book. Moreover, sales of existing books indicated that con-
servatively one could expect to sell 5,000 copies per year
if successful. Actual sales have been about twice this high.
This project was pleasing in prospect and it has been rewarding
in retrospect. This book's revisions will face a more competitive market and give both author and publisher smaller slices
of a much larger pie.

W. H. Freeman and Company has published several books that set
new patterns for what later came to be known as liberal arts
mathematics: books by Sherman Stein (1963), Harold Jacobs (1970),
and Bonnie Averbach and Orin Chein (1980).

In each of these books, matters of content and style are inextricably bound together and each would have vanished without
trace had it not been for the author's and publisher's superb
execution. Each of these books was based on a successful working course and each stood the test of critical reviews before
publication, two criteria that every innovative text should
meet. Each has proved successful in the classrooms of many
other teachers after publication. Each offered a fresh point
of view and a clear alternative for adopters. Each seemed a
sensible commercial gamble, although declining enrollments in
liberal arts mathematics courses in recent years together with
a proliferation of available texts have made this a difficult
market in the 1980's.

Each of these books requires a prepress investment roughly
0.25 times that required for our standard calculus book. For
the earlier books, those by Stein and Jacobs , I estimate
that that investment would be fully recovered with the sale
of 20,000 to 30,000 copies. For the most recent book, that
by Averbach and Chein, the break even point for sales lies
above 37,500 copies and had not yet been reached in 1984.
Compared with a calculus book the investment is about a quarter
and the sales requirement to break even is about half. This
is an interesting sort of gamble. The profit as a portion of
net receipts lies in the range of 10 to 15% for the fully
mature older books. This relatively low rate of return comes
from competitive forces. Long-term survival drives companies
to offer the best product they can at a price low enough to
keep the competition down. The spread between the marginal
cost of printing, paper, and binding and gross receipts is
smaller for books that have many competitors than it is for
those that have few. This is one reason why more advanced
books are more expensive; the relatively smaller size of the
markets for advanced books is another.

Large-scale innovative textbook publishing is so risky that
it becomes attractive only when much of the costs are carried
by others. The committment of the professional community and
of the government to The New Mathematics and to Chem Study
carried forward these two innovations. I cannot cite any
figures for The New Mathematics, but I can cite some for the
initial Chem Study materials published by W. H. Freeman and
Company in 1965. The rate of return on net sales has been
very modest but the net receipts over nineteen years have exceeded eight million dollars and the investment was low. This
chemistry reform project was meticulously developed and backed
by the full faith and credit of the community of chemists and
the government. It was a success in the classroom and spawned
a new generation of successful texts.

The terms of the arrangement allowing W. H. Freeman and Company
to bring out this material forbade the Company from revising
the original Chem Study material and hence from using its position to gain an unfair advantage once the new curriculum was

established. It seems likely that the greatest rewards in
this area were reaped by other publishers who profited by
following the path broken by Chem Study and made improvements
on the prototype published by Freeman.

NEW CURRICULAR MATERIAL AT THE DAWN OF THE AGE OF COMPUTERS

While the debate on continuous versus discrete mathematics
continues, computer technology is creating a new environment.
Students flock to existing discrete mathematics courses at
the sophomore/junior level. These courses are part of the
recommended curriculum of the ACM and they meet, in part, real
needs for computer science departments. How some of this
material can be worked into other courses remains to be seen.
I would hope that the portions of it that are calculus related
(generating functions, expectations, formal expansions, analysis
of certain algorithms, etc.) will show up with more emphasis
in introductory calculus.

Computer technology in the form of TEX, TROF, EQN, and other
typesetting packages, combined with powerful word processing
programs and graphics packages would enable departments to
develop far more polished preliminary editions of experimental
texts. This may allow new books to be produced at substantial
savings. The result is likely to be an unprecedented flourishing of innovative and experimental course materials. The
disadvantage will be that this will increase the amount of
course material to be sorted out by the marketplace. Normally,
publisher plays an important role in sorting out projects and
helping to improve them. As the means of production move more
into the author's hands, this role will be reduced.

The main issue is not whether or not there will be more discrete mathematics in the curriculum, but how the detailed matters will be settled: What discrete-related material will
come to receive more emphasis in existing courses, how new
discrete courses will come to be organized, and over all, how
compelling a mathematics curriculum can be devised and put
into effect. The realization of any change is bound by economic constraints, some of which are outlined here. From my
view it is style and execution that are of central importance

for, as an editor, my job is to find the best author and to
help that author produce the best course materials. In this
context the content is almost a given and the differences in
execution ma   the difference between success and disaster.
Thus the details of style and execution are central for me,
and I represent the editorial decision point. Whatever program is devised must at some point pass editorial judgement
of one sort or another. Here I give you an outline of the
financial constraints on such judgements.

The financial constraints are simply the conditions of life
for publishers. How do publishers determine what projects
are likely to meet their needs? I have given some of the
criteria, but I have left out others that are obvious and
extremely important. Is the project exciting? Is the conception and organization compelling? Is it clear to those within
the company and to reviewers that the author has a real contribution to make? Can the virtues of the product be made evident
to your intended audience? The last is essential to achiev
reasonable sales. We are dealing with course materials, and
so it is natural for potential publishers and for potential
adopters to ask for proof of success in the classroom. The
questions in this area are: Has the material been class tested?
Does it work? Does it appeal to other teachers? Finally we
come to the economic questions: Is there an existing or prospective market large enough to make the project economically
feasible: True, content is part of all of this evaluation,
but success or failure of a book is not usually a matter of
content--content is too obvious a matter. Success comes to
authors with a gift for exposition, who can give the attention
to detail that makes a book work for the author, the author's
students, and for others.

The computer age will bring us new ways to produce books and
new ways to put together instructional materials (courseware).
These developments bring the author closer to the role of com-
positor and allow the author to move into a realm previously
the province of the publisher. This has the potential of
greatly reducing the publisher's prepublication investment in
new teaching material and thus allowing for less costly innova-

tion. This means that the real costs of creating and sorting
out new materials will fall more on the authors (who will collectively produce greater quantities of new materials for limited markets and as a result must, overall, earn a lower average
return) and on the users (who will be confronted by a larger
selection of less-carefully tested products). These developments seem inescapable consequences of present trends.

As these new courses, texts, and other materials are developed,
the content will largely be determined by the perception of
student needs, but successful courses will always depend crit-
ically on the style and presentation of the teacher in the
classroom and of the authors of the materials used. Successful teaching is done in detail, not by choice of content and
large-scale strategy.

REFERENCES

Crawford, James P. 1984. Calculus is not an indescretion.
    The College Mathematics Journal, November 1984.

Hamming, R. W. 1984. Calculus and discrete mathematics.
    The College Mathematics Journal, November 1984.

Kleitman, Daniel J. 1984. Response to Anthony Ralston's
    position. The College Mathematics Journal, November 1984.

Lax, Peter D. 1984. In praise of calculus. The College
    Mathematics Journal, November 1984.

Lucas, William F. 1984. Discrete mathematics courses have
    constituents besides computer scientists. The College
    Mathematics Journal, November 1984.

McMahon, Thomas A., and Bonner, John T. 1983. On Size and
    Life. Scientific American Books.

Wagner, Daniel H. 1984. Calculus versus discrete Mathematics
    in OR applications. The College Mathematics Journal,
    November 1984.

DISCUSSION

NEW CURRICULA AND NEW TOOLS

1. Computers and Computing

2. Calculators

3. Discrete Mathematics

4. Evolution or Revolution ?

STEPS TOWARD A RETHINKING OF THE FOUNDATIONS AND PURPOSES OF INTRODUCTORY CALCULUS

Peter Renz

The concepts and methods of calculus are as important as ever in our understanding of natural science and engineering. One must understand the derivative to work with Newton's laws and one must have a considerably deeper appreciation of the ideas of calculus to begin to work out the bending of beams and the flexing of structural shells. Beyond this, there are many intriguing and important problems before us today that are best initially thought of in terms of interacting continuous variables: the accumulation of pollutants in the ecosphere, the rise and fall of the level of insulin or other medically active material in a patient's bloodstream, the modeling and predicting of the weather, to mention a few. The analysis of discrete processes and of algorithms is often carried out using continuous methods. I mention only the traditional methods of generating functions in discrete probability and recent results on the average efficiency of algorithms to find the zeros of polynomials and to solve linear programming problems. (For a survey of the latter, see Steve Smale [1985].)

How can this be true yet introductory calculus find itself embattled? The answer lies in a set of new realities mainly connected with the microelectronic revolution and the proliferation of calculators, computers, and information processing systems. These devices are finding their way into automobiles, banks, small businesses, private homes, and into our students' bookbags and lives. There are three main ways in which these new realities threaten the traditional calculus curriculum: first, they have led to the proposal that calculus be supplanted by discrete mathematics as a first course in the college curriculum; second, they lead to changes in the sorts of things that can and/or must be taught in calculus itself; third, these new circumstances seem to be associated with a decline in mastery of the traditional background and skills on which the existing calculus curriculum is based. Let us look at these one by one and then see what changes they suggest for calculus.

DEMAND FOR COMPUTER-RELATED MATHEMATICS

The opportunities related to computers and their applications are enormous and consequently students want to learn computer-related material as early as possible in their studies. This demand has led to the proposal that courses in discrete and algorithmic methods be developed as the standard first course in college mathematics, displacing calculus. The pros and cons of this have been argued extensively; see, for example, the Forums on discrete mathematics and the mathematics curriculum with target articles by Anthony Ralston [1984] and Fred Roberts [1984]. The Ralston and Roberts pieces brought forth many persuasive and thoughtful defenses of the central role of calculus with suggestions on what needed to be done better in teaching the subject. I recommend particularly the responses of MacLane [1984], Kleitman [1984], Hamming [1984], Davis [1984], Thomspon [1984], and Guy [1984] and the rebuttals of Ralston and Roberts in The College Mathematics Journal.

These proposals to replace calculus seemed to have been called forth primarily for the strong demand for computer-related courses and this is a reasonable demand that will grow stronger in the future.

CHANGING TECHNOLOGY CHANGES WHAT CAN BE DONE

Computers and calculators have affected the circumstances of introductory calculus in more subtle and complex ways. Essentially all calculus students have scientific calculators. These devices are necessary for working numerical problems in science or engineering courses. Calculators make it possible to obtain, with a few key strokes, answers that previously would have required a great deal of effort and possibly reference to awkward tables. They are the realization of the mythical function machines that were supposed to make the idea of a function concrete when I was an undergraduate. The best device my school could muster then was a square-root Frieden calculator that cost about $1200 in 1957 (about $4500 in 1985 dollars). We have come a long way. That same school is now fully committed to computers in the curriculum, has made microcomputers available to all faculty members, and plans to create a campus-wide network with each undergraduate having his or her own machine. Each of its physics, biology, and mathematics departments already makes extensive use of computers. The broad availability of calculators and computers presents opportunities and difficulties for the teaching of calculus. The new opportunities come because operations that used to be difficult are now easy. Before calculators became almost universal, it would have been unreasonable to ask a student to use iterative methods to solve an equation to eight- or ten-decimal-places accuracy, but now this is a reasonable and rewarding assignment. Similarly, the process of graphing functions and understanding the relationships between a function, and its first and second derivatives and the geometry of the function's graph is transformed when your students can use computer-graphing programs. No longer need you restrict attention to a small collection of functions whose derivatives can be analyzed by simple algebraic means, and quite general functions can be considered. But these opportunities also provide challenges to the traditional calculus curriculum because they substitute new computationally intensive methods for older paper-and-pencil methods. If our students are to understand optimization in a more computational way (look for zeros of f' by Newton's method and check the sign of f"), this implies changes in course content, in the manner that courses are taught, and in the sorts of exercises and tests that are given. These observations are based on my initial experiments using the Microcalc package (Flanders [1985]) in my classes.

CHANGES IN STUDENT PREPAREDNESS

Calculators and computers allow the automation of previously laborious numerical computations, and soon they may routinely deal with symbolic manipulations as well. Perhaps in part because of the lessening importance of calculating skills overall, my students today seem less willing and less able to follow what are essentially computational arguments compared to the students that I taught roughly ten or twenty years ago. As examples I give the standard calculations of the derivatives of x to the nth power or for the sin(x) and cos(x). These may be based on the binomial expansion or the formula for factoring the difference of two nth powers [for the derivatives of x to the nth power] and on the trigonometric angle sum formulas [for sin(x) and cos(x)]. My students today are less familiar with these basic algebraic and trigonometric formulas than were my previous students. Furthermore, today's students also seem less concerned that they be shown a coherent derivation of the results of calculus. Their attitude is that so artfully portrayed by Wilhelm Busch in his cartoon "Firm Faith" reproduced in Figure 1.

Teacher: "... and now I want to prove this theorem."
Pupil: "Why bother to prove it, teacher? I take your word for it."

Professor: „... Und nun will ich Ihnen diesen Lehrsatz jetzt auch beweisen."
Junge: „Wozu beweisen, Herr Professor? Ich glaub' es Ihnen so."

Figure 1.

From *The Mischief Book* by Wilhelm Busch, as translated by Abby Langdon Alger and published by R. Worthington in 1880, reproduced from *Max and Moritz, from the Pen of Wilhelm Busch*, edited and annotated by H. Arthur Klein, Dover Publications, 1962.

They correctly infer that calculus itself must be correct because it seems to have worked well enough for generations of mathematicians, scientists, and engineers, and they see their problem as simply one of mastering the formulas of the subject and how to use them. A demonstration that reduces the problem of finding $\sin'(x)$ to using the addition formula for the sine and to working out $\sin'(0)$ and $\cos'(0)$ is not any help to a student who is not familiar with this addition formula. It is far easier for such a student to simply memorize $\sin'(x) = \cos(x)$. Similar considerations undercut the effectiveness of many proofs traditional to introductory calculus. Consider the chain of reasoning that takes us from, say, completeness of the real numbers to a proof of the mean value theorem. This proof is an exercise in explaining what is geometrically obvious in terms of what is unfamiliar and perhaps almost incomprehensible for many students.

What is needed are compelling ways to make the basic results of calculus directly obvious for such students. For example, the fact that $|\sin(a) - \sin(b)|$ is no larger than $|a - b|$ can be proved by a calculus argument, but isn't it more obvious to observe that, if two points lie at counter-clockwise distances a and b respectively along the circumference of the unit circle starting from point $(1,0)$, then the distance between these two points must be no larger than $|a - b|$, and hence the projections of these points onto the y-axis that given $\sin(a)$ and $\sin(b)$ must satisfy the relation that $|\sin(a) - \sin(b)|$ is no larger than $|a - b|$. A picture here is worth more than a thousand words. This sort of geometric argument showing why sine and cosine are continuous is more evident and persuasive than arguing by reducing the general problem of continuity to that at zero by using the angle sum formulas.

It is not easy to see exactly how this change in level of preparation has come about. It is perhaps partly due to the fact that students have been taking less mathematics in high school, but the students who take calculus at Bard College, where I teach, have had the prerequisite high school courses: algebra I, geometry, and algebra II/trigonometry--although the course coverage for some may have been light in trigonometry. Nonetheless, my students are weak in the concepts and skills that one would expect to be common to such courses. Those skills are outlined by Joan R. Leitzel [1983] in her article which sketches the problem of the decline in mathematical skills of entering college students and suggests solutions, including more clearly defining what should be taught in high school. This lack of skills is reflected not only by classroom experience but also by the MAA's Calculus Readiness Test, on which only a third of our students entering calculus achieved scores at or above the minimum level other schools nationally considered acceptable for entry into calculus. The MAA's testing and placement program which we instituted at Bard in the fall of 1985 gave us the first systematic and quantitative view of the level of mathematical skills and knowledge of our students. It was very helpful not only for those teaching mathematics but also for all those teaching quantitative courses or advising students. At a large school, such as Ohio State where Joan Leitzel teaches, it is possible and necessary to develop a system for placing students on separate tracks and for handling problems of remediation. Smaller schools will not be able to support as elaborate a system as Ohio State. At Bard we are at the opposite end of the scale, with about 750 students and with more of these inclinded to the humanities and arts than to science. We teach a functions/precalculus course once a year in the spring and typically enroll about thirty to forty students in introductory calculus in the fall. Student-scheduling constraints largely determine which calculus section students take, and the weak and strong students are mixed in each class. Under these circumstances, ideally small classes (about fifteen each) allow for individual attention to students but still leave one lecturing to a very mixed audience.

Whatever the causes of these changes and whatever remedies are applied to strengthen high school mathematics, it seems almost certain that tomorrow's calculus students will be less facile with algebra and trigonometry than were the students for whom the traditional calculus course was designed. In partial compensation, powerful scientific calculators and microcomputers will be more common and students will have greater opportunity to observe how mathematical entities behave. For such students the natural logarithm and the common logarithm will be equally easy to deal with. These will simply be two functions that transform products into sums and have rather similar-looking smooth curves as graphs. Both are available at the touch of a calculator key. If those teaching calculus are lucky, their students will have seen logarithms used to scale light or earthquake intensities, star magnitudes, the loudness of sounds, or the concentration of hydrogen ions. If this comes to pass, it will be a reasonable compensation for a somewhat diminished standard of algebraic abilities. For today's and tomorrow's students, there is little need to prove that the natural logarithm function _exists_ by defining that function as an integral, just as there is little need to prove today's students that the sine, cosine, and tangent functions can be defined and their properties rigorously developed starting from the definition of the arc tangent as an integral--a development that was given in detail in the calculus course that I took as a freshman. We may regard the existence of all of these functions as evident. Their properties may be learned most easily by observation and analogy rather than by elaborate proofs.

More generally, we are at a time when machines are being used where paper and pencil and mental muscle used to do all the work. Paper and pencil and mental muscle will never become obsolete, just as arm muscles will never become obsolete. However, today we have no need for the well-muscled armies of laborers who dug ditches and laid tracks in the past. This work is largely done with the aid of machines now. This does not mean that we do not need people who know how to wield a pick and shovel, just that we need for less of them. It is not clear exactly what the mathematical analog of obsolete pick-and-shovel work will be in the future but it already includes arithmetic involving large numbers of digits and all the work that used to be done by looking up logarithms or logarithmic trigonometric functions. In the future, as software like muMath or MACSYMA becomes generally available, it may include much more besides.

THE PURPOSES OF INTRODUCTORY CALCULUS

Calculus has been, is, and will continue to be a basic computational and conceptual tool for students studying the hard sciences and engineering. For most of these purposes a brief sketch of the subject and its methods suffices. Daniel Kleppner and Norman Ramsey [1965] wrote one such sketch, Quick Calculus, for freshman physics students at Harvard. Kleppner and Ramsey wrote their Quick Calculus because the Harvard mathematics department was teaching slow calculus-- that is, a careful and rigorous introduction to the mathematical foundations of calculus, including the completeness of the real numbers, a rigorous treatment of limits, and so on. There is nothing wrong with such slow calculus courses. I took and profited from such a course as a freshman, and I have taught such courses. But I believe that for most students in the age of the calculator and computer, slow calculus is inappropriate and some form of quick calculus or something like a modernized and expanded version of Sylvanus P. Thompson's [1970] classic Calculus Made Easy is more to the point. I believe that this can be true even for students continuing in the mathematical sciences. This is not to downplay the importance of rigor and of careful definitions. We should be careful about what we do and how we do it, but there is no way that a year's course in introductory calculus could cover the usual range of techniques and applications and give the rigorous basis for all the techniques covered. Consider the difficulties of rigorously justifying implicit differentiation, for example, at this level. Slow calculus even has drawbacks for mathematics majors because it delays the introduction of powerful and useful methods until after a full theoretical justification for these methods can be given. Implicit differentiation is again a good example. Surely one wants this available early on. If we accept that the purpose of calculus should be turned toward applications within mathematics and computer science as well as in the sciences and engineering, we are still left with serious questions about what to teach. One view about teaching mathematics for applications is given by Robert Geroch:

> The . . . problem is that it takes a certain amount of effort to learn mathematics. Fortunately, two circumstances here intervene. First, the mathematics one needs for theoretical physics can often be mastered simply by making a sufficient effort. This activity is quite different from, and far more straightforward than, the originality and creativity needed in physics itself. Second, it seems to be the case in practice that

the mathematics one needs in physics is not of a highly sophisticated sort. One
hardly ever uses elaborate theorems or long strings of definitions. Rather, what one
almost always uses, in varous areas of mathematics, is the five or six basic
definitions, some examples to give the definitions life, a few lemmas to relate various
definitions to each other, and a couple of constructions. In short, what one needs
from mathematics is a general idea of what areas of mathematics are available, and in
each area, enough of the flavor of what is going on to feel comfortable. This broad
and largely shallow coverage should in my view be the stuff of "mathematical
physics."

There is, of course, a second, more familiar role of mathematics in physics: that of
solving specific physical problems which have already been formulated
mathematically. This role encompasses such topics as special functions and solutions
of differential equations. This second role has come to dominate the first in the
traditional undergraduate and graduate curricula. My purpose, in part, is to argue
for redressing the balance.

Geroch [1985], p. 2

For Geroch, it is the general ideas and a few illustrative examples that give
the needed sense of the subject. This view is too soft for most engineers.
The proper balance between hard calculations and special functions and
softer, general ideas will be difficult to judge and will vary with the
situation and with the times. Whatever is taught will certainly be largely
forgotten unless it is regularly used, so it is illusory for client engineering
departments to imagine that if hyperbolic functions are taught in
freshman calculus, then their students will have or should have a firm
grasp of these functions when they may have call to use them as juniors.
For this reason it seems desirable to lean in the direction Geroch suggests
rather than to do much work with entities like confluent hypergeometric
functions, for example, a topic that was covered in my sophomore calculus
course and that I have not had call to use since and that is consequently
largely lost to me.

In this view the purposes of introductory calculus are to give the students
an understanding of the main working ideas of the subject:

* The notion of a function made as concrete as possible by practical
examples and graphs.

*The notion that quantities or functions can be effectively defined by
successive approximations precisely when these approximations can be
made as accurate as we please, at least in theory. The first example of this
is the dreaded epsilon-delta defintion of limit. I prefer to think of this
definition as concerned with practical estimation following Hamming [1966
and 1968] rather than as a highly theoretical idea. Other examples include
the definition of the definitie integral and the uses of power series, Taylor
Series and the like.

* The idea of the derivative as a rate of change again with concrete
examples, graphs, and tangent lines to give substance to this idea.

*The first and second derivative as tools for determining the local behavior of a function as used in solving optimization problems initially in one and later in several variables.

*Linearization as used in approximation as in approximating a function by its tangent and in Newton's Method for locating zeros.

*Taylor Series for standard functions including sine, cosine, exp, 1/(1-x), ln, arctan, and Newton's bionomial series. These series are broadly useful and one can show that they converge to their functions by using direct arguments without needing to rely on Cauchy's criterion or completeness if you assume the functions that are being expanded exist.

*An understanding of the sorts of problems in which one must recover f(x) given f'(x) and the applications of indefinte integration (or antidifferentiation) based on formuals for differentiation. The specific techniques to be covered are somewhat dependent on the students' needs and the degree to which this work can be mechanized, but the idea of the process and its usefulness and the knowledge that there are systematic ways to find integrals is essential.

*A unified understanding of definite integration as both a conceptual and computationally effective way to calculate arc lengths, areas, volumes, and other such quantities.

*Sufficient examples and techniques to give life to these ideas and to enable students to use them in other courses. It is important not to give too much in these areas and to find ways to meet the needs of students going on to, say, engineering without rushing them through a large number of special methods none of which is actually mastered. I believe that this can be done by allowing much of the difficult and messy calculation to be done by machine, as is done in practice, while ensuring that the general methods and results are clearly and compellingly evident.

These seem to be a bare minimum of ideas and techniques that should be included in the first year of calculus. What should be found in the third semester of calculus? Here the choices are less clear but the traditional subjects are: geometry and calculus in two and three dimensions including differentiation, optimization problems, integration, and some vector calculus. Some schools add a bit of differential equations to this. These are all fine topics but they do not form a coherent whole that can be easily taught to students who are not at ease with vectors and matrices. For this reason there was an effort supported by CUPM to interpose a linear algebra course between the first year's course in single variable and the traditional third semester course. This idea largely failed to catch on because of the demands of client departments and the inertia of existing programs. Despite this it was a good idea and it seems essential to cut the present monolithic calculus course into pieces if we are to reform it simply because several small jobs are more managable than one immense one. The success of second year several-variable calculus books such as Vector Calculus by Marsden and Tromba [1981], which sells about nine thousand copies a year, shows the possibility of cutting up what has been monolithic. Moreover, from the point of view of course development, text writing, and publishing, the advantages of smaller scale efforts are obvious. See Renz [1985] for an editor and publisher's view on how new courses come into being and how new course materials are developed and published. Thus I restrict my comments to the content of the first year course here.

Any rethinking of introductory calculus must begin with a sharp focus on a limited number of ideas and techniques in order to allow the possibility of significant change. The present course is full to overflowing and improvement cannot be made by addition. Experiments with new curricula must be undertaken with the clear understanding that the needs of client departments in the sciences and engineering must be met but also with the understanding (shared with those clients) that their students will not be needing or using all of the treasured topics of the past. The above is my current list of essential ideas to be covered. I do not expect it to be necessarily adopted by anyone but I would hope that it might lead others to draw up their own lists and to refine them after consultation and to create worthy alternatives to the present calculus courses. One thing that I have purposefully left out of the above is what might form the theoretical basis of such a calculus course and I turn to this question below.

THE FOUNDATIONS OF INTRODUCTORY CALCULUS

Calculus made a good deal of progress based on the somewhat loose ideas of Newton, Leibnitz, and their successors. According to Kline [1972], D'Alembert saw that Newton's ideas of prime and ultimate ratios were the right way to look at the derivative. It was not until Bolzano, Cauchy, Dirichlet, and Weierstrass began to try to put analysis on a firm foundation in the early- to mid- nineteenth century that our modern understanding of calculus began to emerge. I take this foundation to be our understanding of the real numbers as a complete ordered field containing the rational numbers and constructed by considering Dedekind cuts in or Cauchy sequences of rational numbers. To this we must add the set theoretic idea of a function that liberates use from the idea of functions as given by algebraic formulas and the rigorous defintions of limits, and of derivatives in terms of limits, and of integrals either as limits of Riemann sums or in terms of sups and infs. All of this is enough to prove things such as the existence of a real number that is the square root of two--a fact that was never in question for most students--and to show the existence and differentiability of all the elementary transcendental functions. This, however, is all part of a baby real variables course and is not an appropriate theoretical basis for introductory calculus courses.

Every effort should be made to find less complex and subtle assumptions to underlie the methods of calculus taught in the introductory course. We are at liberty in our minds to chose starting points other than completeness or the fact that a continuous function on a closed interval attains its maximum and minimum on that interval (the standard assumption used to derive Rolle's theorem and then the mean value theorem). Why not simply assume as an axiom that the continuous image of a closed interval is a closed interval (giving the existence of maxima and minima and the intermediate value property for continous functions on closed intervals immediately) and why not simply assume the mean value theorem itself? Both of these are geometrically compelling facts whose derivations will be less persuasive than the pictures. This proceedure loses no rigor because what we assume is true is, in any case, actually true for the systems we will work with. All that is really lost is the chance to show a nice and managable proof in class.

If we follow this line, we will look for a small number of geometrically or algebraically evident assumptions on which to base introductory calculus. This will mean going away from the arithmetization of calculus of Cauchy and Weierstrass and back to more geometric ideas--properties of continous and differentiable functions that are easily pictured, such as the intermediate value property and the mean value property for differentiable functions. The elementary transcendental functions can and should be assumed to exist and be differentiable, and their deriatives should be worked out as much as possible by direct calculation using their familiar properties [e.g., by showing that $\log'(x) = (1/x) \log'(0)$] and not by elaborate theoretical means. For example, it is possible to give a direct geometric argument that $\sin'(t) = \cos(t)$ and $\cos'(t) = -\sin(t)$ by considering the motion of the point $P(t) = (\cos(t), \sin(t))$ along the rim of the unit circle. If $P(t)$ advances counter-clockwise as if carried by a wheel whose rim has a linear velocity of one unit per second, then the magnitude of the velocity of $P(t)$ is 1 at all times and its direction is perpendicular to the radius from $(0,0)$ to $P(t)$. It is easy to see that the vertical velocity of $P(t)$ is $\sin'(t)$ and that geometrically this must be $\cos(t)$, and so on. Aided by a physical model I found this derivation a great success with my students this year. This gives an idea of what I feel to be the proper foundation for introductory calculus. To actually work out such a solid yet more intuitive foundation, much careful and imaginative thinking will be needed. But this rethinking will not be enough to lead to successful reform of introductory calculus. What will be needed is many successful experiments. This is a point made in my paper (Renz [1985]) on style versus content in courses and texts.

THE NEED FOR EXTENSIVE AND WELL-SUPPORTED EXPERIMENTAL PROJECTS TO DEVELOP NEW COURSES

There can be no general shift to a new sort of introductory calculus course until there are proven successful models and solid and attractive material from which to teach such a course. To this end small experiments of the most open and imaginative types should be supported. Nothing significant can be accomplished by fine tuning the existing calculus course. It is too constrained, too packed with material. What we need to do is to support sensible but radical experiments. Here support means not only a certain measure of financial support (released time for course development, secretarial support to help produce course notes) but also the more important support that colleagues can give each other of encouraging another in a difficult and chancy enterprise and perhaps even of joining others in the struggle to produce a stronger and more vital calculus course that speaks more directly to today's students and to their present and future needs. The need for new models for calculus courses is evident, the possibilities are there for radical change, but only with extensive experiments properly nurtured can we hope to see this new course evolve.

BIBLIOGRAPHY

Robert B. Davis [1984], "It's Not What You Do, It's How You Do It," The College Mathematics Journal 15 (1984), 391-392.

Harley Flanders [1985], Microcalc, computer software package for introductory calculs, W. H. Freeman and Company Publisher, New York, 1985.

Robert Geroch [1985], Mathematical Physics, University of Chicago Press, Chicago, 1985.

Richard K. Guy [1984], "How Can We Lead in an Up-To-Date and Fair Fashion?" The College Mathematics Journal 15(1984), 396-397.

R. W. Hamming [1966 and 1969], Calculus and the Computer Revolution, published by CUPM, 1966 and by Houghton Mifflin, 1968.

R. W. Hamming [1984], "Calculus and Discrete Mathematics," The College

Mathematics Journal 15 (1984), 388-389.

Morris Klein [1972], Mathematical Thought from Ancient to Modern Times, Oxford University Press, New York, 1972.

Daniel J. Kleitman [1984], "Calculus Defended," The College Mathematics Journal 15(1984), 377-378.

Daniel Kleppner and Norman Ramsey [1965], Quick Calculus, John Wiley and Sons, 1965.

Joan R. Leitzel, "Toward a Common Understanding of the Content of College Prepartory Mathematics," The Two Year College Mathematics Journal 14, (1983), 206-210.

Jerrold E. Marsden and Anthony J. Tromba [1981], Vector Calculus, W. H. Freeman and Company, New York, 1981.

Saunders MacLane [1984], "Calculus as a Discipline," The College Mathematics Journal 15 (1984), 373.

Antony Ralston [1984], "Will Discrete Mathematics Surpass Calculus in Importance?" and Ralston's responses in The College Mathematics Journal 15 (1984), 371-373 and 380-383.

Peter Renz [1985], "Style versus Content: Forces Shaping the Evolution of Textbooks and Courseware," New Directions in Two Year College Mathematics, Donald J. Alberts, ed., Springer Verlag, New York, 1985.

Fred Roberts [1984], "The Introductory Mathematics Curriculum: Misleading, Outdated, and Unfair" and Roberts's response to comments in The College Mathematics Journal 15(1984), 383-385 and 397-399.

Steve Smale [1985], "On the Efficiency of Algorithms of Analysis," Bulletin of the A.M.S. 13 (1985), 87-121.

Patrick W. Thompson [1984], "Issues of Content versus Method," The College Mathematics Journal 15 (1984), 393-395.

Silvannus P. Thompson [1970], Calculus Made Easy, second edition, St. Martin's Press, New York, 1970. Note: The date is deceptive; this book has been in print for over 70 years with little substantial change.

Daniel H. Wagner, "Calculus vs Discrete Mathematics in OR Applications," The College Mathematics Journal 15 (1984), 374-375.

SOME SYSTEMIC WEAKNESSES

AND

THE PLACE OF INTUITION AND APPLICATIONS

IN CALCULUS INSTRUCTION

A Contributed Paper

to the

Sloan Foundation Conference on Calculus Teaching

New Orleans, January, 1986

by

Stephen B. Rodi, Jr.

Austin Community College

Austin, Texas

SOME SYSTEMIC WEAKNESSES

AND

THE PLACE OF INTUITION AND APPLICATIONS

IN CALCULUS INSTRUCTION

by

Stephen B. Rodi, Jr.
Austin Community College
Austin, Texas

In the preliminary planning for this conference, the pasture fenced off for my grazing was "rigor versus intuition" and "the importance and necessity of applications in calculus instruction." There is plenty to feed on within those boundaries but one cannot resist being a curious herbivore and first nibbling a bit at the edge of other fields.

Strong words called this conference into being: the teaching of calculus is in a state of disarray and near crisis at most American colleges and universities, the principal evidence being a failure rate of about 50%. I guess I am not as pessimistic about the general state of affairs as that clarion call suggests I should be, though I agree that the failure rate is too high and that we could be doing a better job.

We should not lose sight of the fact that many good things are going on in calculus instruction. To mention one which I think is often a scapegoat: text material. While calculus books have more topics than can be covered in three semesters and simultaneously weigh and cost too much, the good ones--and there are many--are full of fine features. They carefully, clearly, and systematically lay out material. They are rich with examples and models. They strike a nice balance between the theoretical and the practical. They have a multitude of learning aides. They are colorful, interesting, and graphic.

I have sad memories of the dull, monotonous, confusing, overly abstract text I used 25 years ago to study calculus. To make sure this was not some old nightmare with no basis in fact, I pulled that former

text out again recently.   We are in a land of milk and
honey by comparison.   No better selection of materials
to use in teaching calculus has ever existed.

The Sources of Trouble

So   what has caused all the problems   in   teaching
calculus?   This   observer   comes   to   the   melancholy
conclusion   that   an important part of the problem   lay
not with materials or with students but with the benign
neglect   of undergraduate instruction that has come   to
characterize too many mathematics departments and large
universities in general.

Yes, calculus enrollments are larger and with
larger enrollments come an increased percentage of
underprepared students.   Yes, a wider range of students
(e.g.,   business   and social science majors) are taking
calculus. Yes, over the last two decades our secondary
schools have lowered expectations across the board   for
all students, but especially in mathematics, sending on
to   all   colleges (not just community colleges like my
own)   increasing   numbers of apparently bright young
people who cannot spell, punctuate, factor, or solve a
simple   linear equation.   Particularly   in   our   school
systems, we have become a society where form and
appearance have come to dominate over content and
substance.

These   are important trends in student preparation
that have led to deterioration in many college courses,
not   just   calculus.   But these trends are just now
starting to reverse themselves,   as the nation comes to
realize   that it is producing a generation of certainly
unsophisticated,   and   probably even illiterate,   high
school graduates.   The other guy is starting to   clean
up his act. One only hopes the mess has not become too
large to be unmanageable.

Benign Neglect

But   we at the colleges and universities ought   to
be   concentrating on cleaning up our act.   At the same
time   that all those bad things were happening in high
schools,   we   lived through two decades in   which
university   faculties,   mathematics and   otherwise,
increasingly   disassociated   themselves   from
undergraduate   instruction,   particularly   from

instruction at the freshman-sophomore level. Teaching
such courses became an onus to be shared in at the
minimum level required to keep one's job. There were
no important rewards for teaching them and many
important rewards for directing one's time and energy
elsewhere.

Some of this re-direction of faculty involvement
away from freshman-sophomore instruction is the result
of administrations giving more weight to "research" in
its promotion decisions. But part of it has been a
somewhat solipsistic and selfish decision by
mathematicians and mathematics departments to pursue
their own thing. Sitting around in the afternoon with
a group of 4 or 5 colleagues discussing subtle points
of algebraic topology feels better, is more personally
rewarding, and maybe even easier, than worrying and
struggling about exposing 18 year olds for the first
time to some of history's great ideas like limits and
derivative. No wonder then that departments set up for
themselves systems that miminize faculty hours in the
classroom and in contact with students (like teaching
one section of 120 rather than 3 sections of 40),
systems which enhance most everything else a faculty
member does but diminishes the effectiveness of
freshman-sophomore teaching.

The intention of these remarks is not to indict
but to explicitly recognize the state of affairs. That
recognition must precede change and is the pre-
condition for change. Nothing we discuss at this
conference will have any impact on calculus instruction
unless there is (again as a pre-condition) a dramatic
change in faculty involvement in undergraduate
instruction. We must find ways of attracting to that
enterprise large numbers of perceptive, articulate,
energetic mathematicians who want to work with
undergraduates (particularly freshmen and sophomores)
and guide these young people at a critical juncture in
their personal intellectual growth. Our departments
deliberately have to choose faculty members for these
skills, as systematically and deliberately set out on
this enterprise as they have set out in the past two
decades to build collections of complex analysts or
group theorists. Freshmen and sophomores will not
learn better, calculus instruction will not inprove,
and the conclusions of this conference will not be
efficacious if we continue to have in our classrooms at
that level too high a proportion of people who do not
want to be there and do not have the personal skills to
deal with that kind of communication.

Realism About Students

Another pre-condition for the success of the recommendations of this conference or any other to improve calculus instruction is more realism about student preparation and student readiness to benefit from a calculus class. It was not always the case that students leaving high school and entering a professional program in science, engineering, or business would be expected to go immediately into a calculus course. In fact, proportionately, relatively few did. I went to an advanced placement high school with five years of mathematics and then on to a private liberal arts college as a mathematics major. I took a year-long, thorough course in college algebra, trigonometry and analytical geometry (including rotations and some spherical trig) before beginning calculus. That might have been a little overkill. But I was ready to think about calculus when I got there.

Today, we expect students who at best have three years of high school mathematics of a less sophisticated nature than 25 years ago to go directly into a calculus course. These students simply are not intellectually ready for it. They do not have the skills in algebra and trigonometry. More importantly, they have not had enough exposure to develop the precise and refined mode of thinking required to deal with calculus. It is no wonder a young person who can barely recognize a common factor in a three-term polynomial is befuddled by the careful thinking required to understand the idea of limiting value of a function at a point or to understand why certain ratios are indeterminate and require l'Hopital's rule.

Such students who do not go directly into calculus are labeled in some sense deficient or remedial. No student wants such a label, particularly if the student feels this will retard progress towards his chosen degree. To avoid it, students use every conceivable ruse to get into that calculus course -- like someone angling into French IV who is at best ready for French II. We all know the consequences: either a high drop out rate in the course or an adjustment downward by the instructor in the course's orientation to meet the student's level.

Is there really any hope of significantly improving calculus instruction if we continue to admit--and even indirectly encourage students to seek admission--who are not ready for the course? By far the largest number of high school graduates have three or fewer years of mathematics. This likely will continue to be the case even after the implementation of new higher secondary school standards. Would it not be more consistent to make <u>normal</u> expectation for entering students a thorough pre-calculus course before attempting the heady and adult stuff of calculus? Should we not engage our colleagues in colleges of science, engineering and business in discussions about the realism of building degree plans which <u>expect</u> students to be placed in calculus first semester of freshman year? If we do not ask these questions, do we risk (no matter how perceptive the recommendation of this conference) perpetuating a revolving calculus door through which the brightest safely pass but large numbers of others are trapped because too much has been asked of them too soon?

Human beings have a long developmental period, first in the womb and then in the family, before succeeding alone as adults. Human intellectual growth is also slow (though extraordinary by comparison to other mammals). Granted, the educational system in the U.S. develops mathematical skills less rapidly than possible, as the experience of students in Asia, the Middle East, and Europe demonstrates. Students are theoretically capable of being better prepared mathematically at age 18 than most U.S. students are. But, as a matter of <u>fact</u>, most U.S. students are not. And we build on sand if we attempt significant improvement in calculus instruction without recognizing this.

The Importance of Intuition

I now need to get back to my own pasture before the conference herdsmen discover me with my head over a neighbor's fence. Let's begin grazing back at home with some thoughts on intuition.

To me, the matter is simple and clear: intuition always (or at least almost always) takes precedence over rigor. I will immediately add the disclaimer that I might argue differently for isolated, specialized sections of a calculus course, like an honors section

for mathematics majors.   But such specialized sections
are so rare as not to substantially alter my opinion.
I also will add an additional disclaimer that intuition
for me does not mean "easy" or "inaccurate" or
"incomplete."   Intuition is consistent with requiring
high proficiency in complicated algebra, extensive
knowledge of trigonometry, precision, and thoroughness
in solutions.

What I really mean by intuition is a transmission
to the student (or an educing from the student,
depending on your epistemology) of a sophisticated
<u>understanding</u> of the basic concepts of calculus.   The
first calculus course should have as one of its major
goals sending forth a student with a substantial
comprehension of ideas like limit, derivative,
integral, infinite process (both large and small), and
with an accurate knowledge how these ideas are
pertinent to the solution of a whole range of important
problems.   Understanding, not regurgitation, is the
goal.

I sometimes say to my students that I think I
could teach Koko, the push-button, talking gorilla, the
mechanics of the use of l'Hopital's rule, given enough
time and effort.   What I am not optimistic about are
Koko's chances of understanding the nature of an
indeterminate form.

I deliberately choose l'Hopital's rule as an
example since, every time I teach it, it strikes me as
an especially good example of what intuitive,
conceptual teaching is <u>not</u>. Every textbook states and
proves the relevant theorems and give a plethora of
examples of l'Hopital variants.   I have yet to see one
text which explains what "indeterminant" <u>means</u>.
Students become masters of the mechanics of the process
but have absolutely no understanding of the basic
problem: that quantities apparently similar because
they simultaneously grow large in the numerator and
denominator may have ratios with very different
behavior. Students cannot explain that "indeterminacy"
is precisely a state in which knowledge about the
(separate and independent) large growth in numerator
and denominator does not by itself resolve what is
happening to the ratio.   Further information is
required, which one can get from l'Hopital.

As an aside, I will note that this is an ideal topic with which to use the computer to produce insight. It takes two minutes to display three different ratios, each with numerator and denominator having infinite growth, but the ratios themselves either tending to zero or to some other finite limit or growing without bound. A single glance at these examples convinces students of the "intellectual point" of l'Hopital. Absent these examples, most students go through the section like Koko.

There are many other places where intuition in the sense of conceptual understanding is the key. I will say more about this below when discussing applications. For now, I want to emphasize that I am talking about the frame of mind with which the instructor presents the course. Neither the simplistic presentation of calculus as a set of routine techniques (e.g., rules of differentiation) nor the rigorous development of theorem after theorem each in its full panoply of epsilons and deltas catches this "intuitive" spirit. The instructor has to be constantly asking "What is the underlying concept here?", "What is the problem being attacked?", "What is the generalization that carries over to other situations?"

Student at the level of calculus do not make these connections and see these patterns for themselves. They have had very little opportunity to do so. They are only beginning to realize the possibility of such sophisticated thinking, in terms of both concept content, logic, and generalization. The duty of the instructor is to start the student on this road and give the student substantial guidance in the early part of the journey.

One obvious corollary (at least for me) of leading the intuitive life in calculus is to be concrete, which also means to be visual. Use pictures of various kinds of discontinuities, show computer printouts of limits, display three-dimensional models of surfaces in space. Most of these activities can take place in just a few minutes in class, but they make a massive difference in generating understanding in neophytes. And they are just the sort of thing a hoarier mathematician, who has become accustomed to life in n-space where the rules of process are more important than the visual image, might forget about.

The scholastic philosophers argued <u>nihil est in intellectu nisi prior in sensibus</u>, nothing is in the intellect which is not first in the senses.

An intuitive approach to calculus is not inconsistent with detailed and complicated examples. Intuitive does not mean trivial. The ideas that students are beginning to understand in this course are complex. As a result, many and complex examples and models will be needed to develop them. In fact, they are probably best developed by seeing them time after time in a cyclic process, not unlike the exposition most calculus textbooks now follow: first developing polynomial functions; then rational functions; then in turn logarithmic, exponential, trigonometric, inverse trigonometric, and hyperbolic functions. At each stage, one learns critical information about the new model as a tool to revealing facets of the keystone ideas that hold all of calculus together.

Applications

This brings me to the second half of my assigned pasture: applications. Applications are a critical part of teaching and learning calculus precisely because they constitute one of the best places both to expose and to reinforce intuitive conceptual understanding.

A real danger when research mathematicians teach calculus to freshman and sophomores is forward projection: the tendency for the mathematician to look ahead to the next generalization (the class of Riemann integrable functions as a precursor of Lebesque integration) rather than look back to the kind of concrete problems which gave rise to calculus in the first place and which are necessary for student understanding. Insistence on dealing with practical applications purposely woven into the course can offset that bias.

The bias sometimes is so strong that almost no applications are done in the calculus course. The philosophy in some departments apparently is that the physicists and the engineers do applications in their courses; mathematicians only to "mathematics." The irony is that the physicists and the engineers "assume" the mathematicians have done the job. The losers are the students.

The key word is using applications in a   calculus
course is "purposeful." I suspect no one would attempt
to   teach the definite integral without   proposing   the
question of finding the area "under" a   curve.   Why?
Because that geometric model simultaneously motivates
the   consideration of definite integrals in   the   first
place and constitutes a dramatic picture of how and why
the   sum is constructed as it is.   Such a presentation
gives insight to a new idea in a way a mere abstraction
like "limit of a certain sum of products" never can.

But, unfortunately, the spirit of that example
generally is not carried systematically through the
entire calculus course.   The real principle underlying
that first presentation of the definite integral in the
ancient Greek one of learning about the irregular
through the regular or of extending and adjusting the
known to analyze and reveal the unknown.   (For the
Greeks, the area of a circle became the limiting case
of   inscribed   regular   polygons.)   Those   ancient
principles,   monumental   steps   in   the   historical
development of problem solving and still valuable tools
today, can be re-taught time after time in calculus, if
the instructor elects to do so .

One does not have to choose a multitude of
examples using the integral to develop such a   theme.
One only needs to choose them shrewdly. And especially
to present them in such a way as to elucidate the
common underlying pattern.   One of my favorites is
water pressure on a submerged plate. This model has an
intuitive appeal to students from almost all majors.
It requires very little knowledge about physics. It is
a marvelous example of mastering the   forces   on   an
irregualr shape (the plate) by partitioning   into
regular rectangular shapes.   Finally, one cannot just
"memorize" an integral formula to solve all such
problems.   The process needs to be understood since
small modifications in the positioning of the plate can
affect the exact form of the integrand.

If one rule is to pick applications that enrich
intuition and display basic processes, another rule is
to use applications that weave their way throughout the
course, allowing students to see each successive stage
of generalization. Moments and centroids are one such.
The topic begins with the simplicity of point masses in
one dimension (idealized children sitting on a teeter-
totter) and concludes with three dimensional solids of
irregular density.   The story in-between is one of the

prettiest in calculus, as one proceeds successively and concomitantly from ordinary addition to triple integrals, from one dimension to three, and from homogeneous sheets to non-homogeneous solids. Teachers need to unpetal ideas to students through applications like this, as though unpetaling a rose, to reveal the more complex inner structure. An imageless poem is a failure; and even worse is a poem full of pointless or contradictory images. Applications are the imagery of calculus, which (if well selected and expressed) are as powerful a tool in revealing its message as metaphor is to the poet.

In choosing and presenting applications, an instructor's emphasis needs to be on what is common, not what is different; what is central, not what is peripheral.

I find students get lost in chapters on local extrema, like spelunkers who wander from cavern to cavern when the path out is simple and straight. So few books make the observation that there are only <u>two</u> underlying intuitive ideas when discussing extrema: increasing/decreasing and concavity. Each in its own way gives the necessary insight to resolve a critical point's status. The first derivative test is just a technical expression of the former and the second derivative test does the same for the latter. Everything else is the mechanics of organizing information or algebraic detail or exceptional cases. It is easy to follow a path in this maze, if one is allowed to see the real simplicity of the design.

One criticism of calculus textbooks is that they grow too long because they contain too many applications. I partly agree and partly disagree. Individual applications are superfluous to the extent that they are addenda, mere appendages hanging between the covers so that the potential adopter can see his or her favorite word in the index. But the technique of using application to teach the subject is not superfluous. Authors might consider putting some applications in an appendix or in a supplemental pamphlet as a way of keeping the main development more coherent and still allowing users some topics to select cafeteria-style. I would also add that authors of calculus texts used primarily by business and social science majors have a particularly challenging task (in light of my comments) since at least to me the

applications in those areas always seem more artificial
(if not sometimes juvenile), making them weak carriers
of the main themes of calculus.

Pruning the Calculus Course

I began this paper by grazing a bit across my
fence line. I will end it by doing the same.

With increasing frequency, one hears that calculus
courses need to be cut back to provide room in the
curriculum for other mathematics courses. I agree that
space has to be found for other topics, particularly
from discrete mathematics and particularly for
undergraduate majors in business, operations research,
and computer science. But I want to caution against a
hasty pruning of that giant oak called calculus. We
may need more trees in the garden, alternate sequences
of courses for students whose needs are different. But
I am not sure we need radical surgery on the ones
already there.

For example, I would encourage computer science
majors to take only two semesters of calculus (calculus
in the plane), using the third semester to develop
discrete topics. But I would want to see physicists
and engineers go on to take a traditional third
semester of multivariate calculus.

Particularly pernicious, I think, are
recommendations to remove almost all computations from
calculus since (the argument goes) software packages
are already available to do them. For example, let all
definite integrals be computed by using numerical
methods on a pre-programmed hand-held calculator.

I am happy that slide rules have become extinct,
one piece of technology replaced by a better one. I am
not happy that neither children in grammar school nor
clerks in department stores can no longer add. That is
a dimunition of basic human understanding unexpectedly
resulting from technology. A colleague of mine on
temporary leave from a large governmental research
laboratory tells of his associates who run to the
computer to get approximate answers to simple first
order linear differential equations that could be
solved by separation of variables and simple straightforward integration.

In the rush to pre-packaged programs, we need to be careful not to toss out the baby with the bath water. In devoting two or three classes to methods of integration, one is doing far more than developing a mechanical skill. The real, perduring learning here is the analysis of patterns, the recognition of the structure of integrands, the ability to choose the proper tool (be it substitution or partial fractions) to attack the problem. The same analytical skill is exercised again when one chooses among a myriad of convergence tests to use on an infinite series.

Let calculators work out trivial details, like the monotonous inversion of large size matrices. But let us not lop off important limbs on the calculus tree, topics which (if taught properly) play a special role in a student's intellectual growth. Once removed, it is difficult to find substitutes and certain skills will atrophy. Calculators need be used as assistants, not as replacements for analysis and thought.

Another danger I see, again arising from a desire for a more compact calculus course to free up time elsewhere in the curriculum, is the production of pre-packaged calculus sequences, a sort of highlights of calculus, like the highlights of art history in a one-semester survey. I was once told that learning philosophy was like climbing a mountain. The same is true for calculus. One can be placed at the top by helicopter, look down and see various paths (including the most efficient one to the top), and appreciate the beauty of the landscape. But this tourist-eye view is far different from having to climb that mountain, correcting false turns as one goes, getting partial insights as you look back, and finally seeing a panaroma which has truly become part of the climber. However we improve the teaching of calculus, we need to leave both the time and the challenge for substantial individual climbing. Only that way does real intellectual growth take place.

Calculus Articulation between High School and College

Don Small
Department of Mathematics
Colby College
Waterville, Maine 04901

One of the most important and difficult problems facing many college mathematics departments today is the development of programs that enable students who have studied calculus in high school to continue their accelerated mathematics program into college. It is important because most of the more promising and capable mathematics students in high school are in accelerated programs and it is difficult because of the wide diversity of precalculus and calculus programs. This paper will analyze and present recommendations on the articulation between high school and college calculus courses. Three areas are considered:

Accelerated high school mathematics programs.

Fifth year programs.

College programs.

I. Accelerated High School Programs

Accelerated mathematics programs beginning with algebra in the eighth grade are now well established and accepted in most school systems. The success of these programs in attracting the more mathematically capable students was documented in the 1981-82 testing that was done with the (twenty nation) "Second International Mathematics Study." The Summary Report [9] states, with reference to a comparison between precalculus and calculus students in the United States:

We note furthermore that in every content area (sets and relations, number systems, algebra, geometry, elementary functions/calculus, probability and statistics, finite mathematics) the end of the year average achievement of the precalculus classes was less (and in many cases considerably less) than the beginning of the year achievement of the calculus students.

The report continues:

It is important to observe that the great majority of U.S. senior high school students in fourth and fifth year mathematics classes (that is, those in precalculus classes) had an average performance level that was at or below that of the lower 25 percent of the countries. The end-of-year performance of the students in the calculus classes was at or near the international means for the various content areas, with the exception of geometry. Here U.S. performance was below the international average.

Thus the students in accelerated programs culminating in a calculus course performed near the international mean level while their classmates in (non accelerated) programs culminating in precalculus courses performed at or below the lower 25 percent level in this international survey.

II. Fifth Year Programs

The success of the accelerated programs in having students complete the standard four year college preparatory mathematics program by the end of the eleventh grade presents schools with both an opportunity and a challenge for fifth year programs. There are two acceptable options for a fifth year program:

1. Offer college level mathematics courses that will continue the students' accelerated program and thus provide exemption from one or two semesters of college mathematics,

2. Offer high school mathematics courses that will broaden and strengthen a student's background and understanding of precollege mathematics.

Not offering a fifth year course or offering a "watered-down" college level course with no expectation of students earning advanced placement or offering a course that just "samples" topics from a college course are not considered to be acceptable fifth year alternatives.

A great deal of prestige is associated with offering calculus as a fifth year program. Communities, in general, often view the offering of calculus in their high school as an indication of a quality educational program. Parents, School Board officials, counselors, and school administrators often demonstrate a competitive type pride in their school's offering of calculus. This prestige factor can easily manifest itself in strong political pressure for a school to offer calculus without regard to the qualifications of teachers or students. It is important that this political pressure be resisted and that the choice of a fifth year program be made by the mathematics faculty of the local school and be made on the basis of the interest and qualifications of the mathematics faculty and the quality and number of available students. School officials should be encouraged to develop public awareness programs to spread the prestige and support for the calculus, to acceleration programs in general. This would help diffuse the political pressure as well as broaden school support within the community.

If a school elects to offer a college level course, then the course should be based on a standard college course syllabus (e.g. The Advanced Placement syllabus for calculus). Furthermore, the evaluation of the course should be based primarily on the performance of its graduates in the next level calculus course.

High School Calculus

There are many valid reasons why a fifth year program should include a calculus course.

1. Calculus is generally recognized as the beginning course in a college mathematics program.

2. There exists a (nationally accepted) syllabus.

3. The Advanced Placement program offers a nationwide mechanism for obtaining advanced placement.

4. There is usually strong community and school support for a calculus program.

A calculus course, however, should not be offered unless there are strong indications that the course will be successful.

Successful Calculus Course

The primary characteristics of a successful high school calculus course are:

1. Qualified Instructor: A degree in mathematics that includes at least one course in advanced real analysis is strongly recommended for anyone teaching calculus.

2. A full year course based on the Advanced Placement syllabus.

3. Use of a college text rather than a "watered-down" high school version.

4. A course whose major goal is to prepare students for advanced placement and not merely to get students ready to repeat calculus.

5. Course evaluation is based primarily on the the performance of its graduates in the next higher level calculus course.

6. Course enrollment is restricted to only qualified and interested students.

7. There exists an alternative fifth year course that the less prepared or less interested students may elect.

The bottom line of what makes a high school calculus course successful is no surprise to anyone. A qualified teacher with high but realistic expectations, using somewhat standard course objectives with students who are willing and able to learn, results in a successful transition at any level of our educational process. When any of the above ingredients are missing, problems appear.

Unsuccessful Calculus Courses

There are two types of calculus courses given in high schools that seem to have become widespread and that have an undesirable impact on students who later take calculus in college. One type is a one semester or partial year course that presents the highlights of calculus, including an intuitive look at the main concepts and a few applications, and makes no pretense about being a complete course in the subject. The motivation for offering a course of this kind is the *misguided* idea that it prepares students for a *real* course in college. However, such a *preview* course covers only the glory and thus takes the excitement of calculus away from the college course without adequately preparing students for the hard work and occasional drudgery needed to understand concepts and master technical skills. Professor Sherbert has commented: *It is like showing a ten minute highlights film of a baseball game, including the final score, and then forcing the viewer to watch the entire game from the beginning - with a quiz after each inning.* Or, as one college placement director said, the short preview course only succeeds in taking the *bloom off the rose.*

The other type of course is a year long, semi-serious, but watered down treatment of calculus that does not deal in depth with the concepts, covers no proofs or rigorous derivations, and mostly stresses computation. The lack of high standards and emphasis on understanding, dangerously misleads students into thinking they know more than they really do. In this case, not only is the excitement taken away, but an unfounded feeling of subject mastery is fostered that can lead to serious problems in college calculus courses. Students can receive respectable grades in a course of this type, yet have only a slight chance of passing an Advanced Placement examination or a college administered proficiency examination. Those who place into second term calculus in college, will find themselves in heavy competition with better prepared classmates. Those who select (or are selected) to repeat first term calculus believe they know more than they do, and the motivation and willingness to learn the subject are lacking.

III. College Programs

In the decade between 1973 and 1982, the number of students studying calculus in high school grew at a rate of more than 10% annually. In 1982, 234,000 students passed a high school calculus course and 148,600 of them received a grade of B^- or higher [2]. Extrapolating these figures to 1985, clearly shows that more than a third of the 500,000 students who took their first college calculus course in 1985 had previously received a grade of B^- or higher in a high school calculus course. Of the 48,351 students who took a 1985 Advanced Placement examination in mathematics, 17,494 received a score of 4 or 5 and 12,296 received a score of 3.

Several studies ([3], [5], [6]) have been conducted on the performance in later courses by students who have received advanced placement (and possibly college credit) by virtue of their scores on Advanced Placement exams. The studies show that, overall, students earning a score of 4 or 5 on either the AB or BC Advanced Placement exam do as well or better than the students who have taken all their calculus in college. It is therefore strongly recommended that colleges recognize the validity of the Advanced Placement program by granting a one semester advanced placement with credit for students with a 4 or 5 score on the AB exam and a two semester advanced placement with credit

There is no clear consensus concerning performance in subsequent calculus courses by students who have scored a 3 on an Advanced Placement examination. The treatment of these students is a very important articulation question since approximately one third of all students who take an Advanced Placement examination are in this group and many of them are quite mathematically capable. It is therefore recommended that students receiving a a 3 on an Advanced Placement examination be dealt with on a special basis in a manner that is appropriate for the institution involved. For example, several colleges offer such a student an opportunity to "upgrade" his or her score to an "equivalent 4" by doing sufficiently well on a Department of Mathematics placement examination. Some institutions give such students one semester of advanced placement with credit for Calculus I upon successful completion of Calculus II. A third option is to give one semester of advanced placement with credit for Calculus I and provide a special section of Calculus II for such students.

Other important articulation problems are associated with students who have studied calculus in high school, but have not attained advanced placement either through the Advanced Placement Calculus program or through effective college procedures. These students pose a serious and difficult challenge to college mathematics departments, namely: How to place these students so that they can benefit from their accelerated high school program and not succumb to the negative and (academically) destructive attitude problems that often result when a student repeats a course in which success has already been experienced? There are three major factors to consider with respect to these students.

1. Lack of uniformity of high school calculus courses. The wide diversity in the backgrounds of the students necessitates a large review component be included in their first college calculus course to guarantee the necessary foundation for future courses.

2. The mistaken belief of many students that they *really know the calculus* when in fact, they do not. Thus they do not take studying seriously enough at the beginning of the course. When they realize their mistake (if they do), it is often too late. These students often become discouraged and

resentful as a result of their poor performance in college calculus, and believe that it is the college course that must be at fault.

3. The *Pecking Order* syndrome. The better the student, the more upsetting are the understandable feelings of uncertainity about one's position relative to the others in the class. Although this is a common problem for all college freshman, it is compounded when the student appears to be repeating a course in which success had been achieved the preceding year. This promotes the feelings of anxiety and produces an accompanying set of excuses if the student does not do at least as well as in the previous year. The uncertainty of one's position relative to the rest of the class often manifests itself in the student not asking questions or discussing in (or out of) class for fear of appearing *dumb*. This is in marked contrast to the highly confident high school senior whose questions and discussions were major components in his or her learning process.

The unpleasant fact is that the majority of students who have taken calculus in high school and have not clearly earned advanced placement do not *fit* in either the standard Calculus I or Calculus II course. The students do not have the level of mastery of Calculus I topics to be successful if placed in Calculus II and are often doomed by the attitude problems if placed in Calculus I. This is the *Scylla and Charybdis* of the articulation problem or, in modern parlance, it's the *rock and hard place.*

An additional factor to consider is the negative effect that a group of students who are repeating most of the content of calculus I has on the rest of the class as well as on the level of the instructor's presentations.

What is needed are courses designed especially for students who have taken calculus in high school and have not scored a 4 or 5 on an Advanced Placement exam. These courses need to be designed so that they:

1. acknowledge and build on the high school experiences of the students,

2. provide necessary review opportunities to ensure an acceptable level of understanding of calculus I topics, and

3. are **clearly different** from high school calculus courses (in order that students do not feel that they are essentially just repeating their high school course),

An additional desirable boundary condition for these courses is that they result in the equivalent of a one course advanced placement.

Altering the *standard* lecture format or rearranging content seem to be two promising approaches to developing courses that will satisfy the above criteria (E.g. Colby College has successfully developed a two semester course that integrates multivariable with single variable calculus [10]. This course satisfies the three conditions listed above as well as providing students with the equivalent of a one course advanced placement.)

Colleges have an opportunity and responsibility to develop and foster communication with high schools. In particular, there are three areas that colleges should develop.

1. Establish periodic (e.g. once a semester) articulation meetings where high school and college teachers meet and discuss expectations, requirements, and student performance.

2. Develop a system for providing feedback to high school teachers on the performance of their students. This would aid the high school teachers in evaluating their courses.

3. Develop, in conjunction with Regional or State Supervisors, enrichment programs in elementary calculus (and other fifth year courses) for high school teachers. These programs could be offered during the summer or during the school year. Institutes, short courses, mini courses, and Chautauqua programs offer models for such programs.

IV. Recommendations

1. School administrators should develop a public awareness program with the objective of extending the support for fifth year calculus courses to accelerated programs including all of the fifth year options.

2. A fifth year program should offer a student a choice of courses (not just calculus).

3. The choice of fifth year programs should be made by the mathematics faculty on the basis of interest and qualifications of the faculty and the quality and number of the accelerated students.

4. If a fifth year course is a college course, then it should be treated as a college level course (text, syllabus, rigor).

5. A college level fifth year course should be taught with the expectation that successful graduates (B^- or better) would not repeat the course in college.

6. A fifth year program should provide a *bail out* option for the student who is not qualified or interested in continuing in an accelerated program.

7. A mathematics degree that includes at least one course in advanced real analysis is strongly recommended for anyone teaching calculus.

8. A high school calculus course should be a full year course based on the Advanced Placement syllabus.

9. High school calculus students should take either the AB or BC Advanced Placement exam.

10. The evaluation of a high school calculus course should be based primarily on the performance of its graduates in the next level calculus course.

11. Only interested students who have successfully completed the standard four year college preparatory program in mathematics should be permitted to take a high school calculus course.

12. Colleges should grant credit and advanced placement out of Calculus I for students with a 4 or 5 score on the AB exam and credit and placement out of Calculus II for students with a 4 or 5 score on the BC exam. Colleges should develop procedures for providing special treatment for students who have earned a score of 3 on an Advanced Placement examination.

13. Colleges should individualize as much as possible the advising and placement of students who have taken calculus in high school. Placement test scores and personal interviews should be used in determining the placement of these students.

14. Colleges should develop special courses in calculus for students who have been successful in accelerated programs and have clearly not earned advanced placement.

15. Colleges should develop communication channels with high schools: articulation meetings, performance feedback, and instructional enrichment.

References

[1] C. Cahow, N. Christensen, J. Gregg, E. Nathans, H. Strobel, G. Williams, Undergraduate Faculty Council of Arts and Sciences Committee on Curriculum Subcommittee on Advanced Placement, Report, *Trinity College, Duke University*, (1979)

[2] C. Dennis Carroll, High School and Beyond Tabulation: Mathematics Courses Taken by 1980 High School Sophomores Who Graduated in 1982, *National Center for Educational Statistics*, April (1984) (LSB 84-4-3)

[3] P. C. Chamberlain, R. C. Pugh, J. Schellhammer, Does Advanced Placement Continue Throughout the Undergraduate Years? *College and University* (winter 1968) 195-200

[4] *The College Board, Advanced Placement Course Description, Mathematics* 1986

[5] E. Dickey, A Study Comparing Advanced Placement and First-Year College Calculus Students on a Calculus Achievement Test, *Ed.D. dissertation*, University of South Carolina (1982)

[6] D. A. Frisbe, Comparison of Course Performance of APP and Non-APP Calculus Students, *Research Memorandum No 207, University of Illinois*, September 1980.

[7] D. Fry, A Comparison of the College Performance in Calculus-Level Mathematics Courses Between Regular-Progress Students and Advanced Placement Students, *Ed.D. dissertation*, Temple University (1973)

[8] C. Jones, J. Kenelly, D. Kreider, The Advanced Placement Program in Mathematics-Update 1975, *Mathematics Teacher* (1975)

[9] Second International Mathematics Study Summary Report for the United States (1985) Stipes Pub. Co., Champaign, IL.

[10] D. Small, J. Hosack, K. Lane, Calculus of One and Several Variables: An Integrated Approach, Colby College (1984)

[11] D. H.Sorge and G. H. Wheatly, Calculus in High School-at What Cost? *American Mathematical Monthly*, vol 84 (1977) 644-647.

[12] D. M. Spresser, Placement of the first college course, *International Journal Mathematics Education, Science, and Technology*, (1979) vol 10, no. 4, 593-600.

Computer Algebra Systems, Tools For Reforming Calculus Instruction

by
Donald B. Small and John M. Hosack
Department of Mathematics
Colby College, Waterville, Maine 04901

I. Introduction

Computer Algebra Systems (CAS), which are becoming increasingly available, have the potential for significantly improving calculus instruction. A major effort on the part of many people to consider issues of pedagogy and content will be required to realize this potential. We hope that this paper contributes to that effort.

The first portion of this paper provides a brief explanation of computer algebra systems. In the main section we describe four areas in which we feel the present instruction in calculus needs to be reformed and describe how the use of CAS could be a major tool in a reformulation. The final section contains responses to three commonly raised questions/objections concerning CAS.

II. What are CAS

Computer Algebra Systems are computer systems for the exact solution of problems in symbolic form. This contrasts with the numerical analysis approach used in conventional computer languages such as FORTRAN or BASIC, where a numerical approximation is obtained. The ones of interest to us (MACSYMA [3], Maple [1], muMATH [4], REDUCE [6], SMP [5]) are interactive systems that allow the user to define an expression, apply an operation, and manipulate the output.

For example, if the user wishes to integrate

$f(x) = \arcsin(x)/x^2$ with respect to x,

the user defines the expression (where " ^ " is exponentiation)

f : = arcsin(x)/x ^ 2

and applies the integration operator

integrate (f,x)

obtaining as output an indefinite integral, usually in two dimensional format:

$$\frac{-log(2\sqrt{1-x^2}+2)}{|x|} - \frac{\arcsin(x)}{x}$$

The standard operations include the use of the system as an arbitrary precision desk calculator, algebraic simplification, calculus (differentiation, integration, power series), linear algebra, systems of equations, differential equations, etc. There are also utility programs for expression manipulation, such as editing expressions and extracting parts of expressions. The systems may also allow numerical procedures such as numerical integration or graphing. If a procedure is not provided, these systems include high-level programming capabilities to allow for user-written procedures as extensions to the system. However, it is important to understand that computer algebra systems do not require any programming on the part of the user for most applications. These systems are readily available to the "computer novice" with a 10 or 15 minute introduction starting with how to turn on the computer. In addition, many of thses systems have on-line help, so that almost no class time need be spent on the mechanics of using the system.

III. Areas for Improving Calculus Instruction

(1) *Conceptual Understanding*

Too much of the time of mathematics undergraduates is spent carrying out routine algorithmic manipulations (which the students will not long remember). This is done at the expense of conceptual understanding of the material and an appreciation of mathematical processes (e.g. problem solving approaches and development of algorithms). Today, the major emphasis in college mathematics instruction is placed on imparting specific mathematical facts and algorithms, rather than on understanding and the development of inquisitive attitudes, analytical abilities, and problem solving skills. Paul Halmos argues [2]

> *The major part of every meaningful life is the solution of problems; a considerable part of the professional life of technicians, engineers, scientists, etc., is the solution of mathematical problems. It is the duty of all teachers, and of teachers of mathematics in particular, to expose their students to problems much more than to facts.*

Since CAS are becoming widely available that can carry out most of the standard operations of calculus, we ask: Rather than emphasizing the blind carrying out of algorithms, which can be done better by CAS, how can CAS be used to enhance conceptual understanding and the development of analytical thinking skills? We offer five illustrations as a partial answer to this question.

(a) Problem Solving Approaches

Shifting the burden of computation to CAS makes time available for students to concentrate on how to approach a problem, to delineate subproblems, and to consider alternatives, rather than spending most of the time routinely following one algorithm. For example, one can develop the basic approach toward problem

solving outlined by the following questions:

1. What do I need to know?

2. What can I tell by inspection?

3. What are the possibilities suggested by what I know?

4. What should I do next?

The value of this approach is difficult to appreciate when the algebra involved can

be routinely done with pencil and paper. This is the case in sketching the graphs

of rational functions when one is only concerned with polynomials of degree 2 or

3, which is typical of most exercises. However, the value of this approach becomes

clear to students when the algebra involved is too complicated to be done with

pencil and paper. For example, sketching the graph of a rational function with

polynomials of degree 8 or 13 or 20.

(b) Increase in the Quantity and Variety of Exercises

One of the tenets of mathematics instruction seems to be that the more exam-

ples students work or see worked, the better they understand the concepts involved.

If this is in fact true, then CAS can be the answer to *drill* work (students get tired,

computers do not). For example, in discussing Taylor series we could ask students

to analyze ten or more functions during a homework assignment rather than just

three or four. Furthermore we would not be limited to only considering the few

standard functions. Using CAS, for example, we could ask students to find a polynomial that would approximate $f(x) = e^{-x^2} cos(x)$ over the interval [-1,4] with an accuracy of 10^{-8}.

(c) Use of Graphs to Analyze Functions

We use the analysis of functions to learn how to sketch graphs, but we do not place very much emphasis on using graphs to help analyze functions. The graphing capabilities of CAS could be used to *guide* the analysis of a function. This could be particularly helpful in extrema problems of one or more variables. For example, a sketch of the fourth derivative might suggest a possible bound to use in the error expression of Simpson's rule. Other examples where sketches could be used to suggest or guide the analysis include showing the convergence or divergence of a sequence of partial sums or comparing two functions or solving an equation (a sketch could approximate a zero and then the bisection procedure could be used, with the computation carried out by the CAS, to obtain the desired accuracy).

(d) Development of an Inquisitive and Experimental Attitude

A major obstacle to developing an inquisitive approach in students is the large amount of (routine) computation that is usually involved in attempting to answer or even understand most questions. With CAS to carry out the computations, it is easy to follow up an exercise with a sequence of "What if ..." type questions in which some aspect of the original problem is changed and we want to know what effect the alteration has on the solution. For example, consider the level of class involvement and the "discovery setting" that an instructor could orchestrate in a classroom equipped with a microcomputer (with CAS) and a projector by altering the integral $\int \sqrt{1-x^2}dx$. By removing or lessening the burden of computation, the instructor can better focus (in class) on developing a conjecturing and experimental approach on the part of students. CAS can make *discovery* learning a real possibility for many students.

(e) Changing Students' Perception of What is Important in Mathematics

Time spent on an activity is often viewed as an indication of the importance of that activity. Since students spend most of their "mathematics time" carrying out routine algorithmic manipulations, their tendency is to view mathematics as the memorization of formulas and "to do mathematics" is to compute. ("I can do the math, it's just the theorem that I don't understand.") A vicious cycle emerges, for trying harder to learn mathematics means putting more emphasis on computing and even less on understanding concepts. Relegating computation to CAS can free the student to think about what is going on, to anticipate the answer (type, form, size, etc.), and to conjecture. In this way, the student is more likely to become involved with conceptual understanding rather than just the details of computation. The use of CAS will not identify what is important in mathematics, but it will downgrade the importance of computation in the students' minds and provide the instructor with a better chance of convincing the student of what is important and what is not so important.

(2) *Approximation and Error Bound Analysis*

An important goal of a college mathematics program should be to prepare students to be able to apply their mathematical education to "real-life" problem situations. The open-ended nature of most of these situations requires an analysis that depends on an approximation approach and error bound analysis. Preparation for this type of analysis belongs in our calculus courses.

Approximation should be considered as the primary process throughout calculus. A two-to-one ratio of open-ended problems to closed-form problems is desirable. For example, twice as much time should be spent on numerical integration as closed form integration; twice as much time should be spent on approximating functions with polynomials as on convergence tests. Error bound analysis should be stressed throughout the calculus sequence. What is meant by "bounding a function" and "how to determine a bound for a function" are important considerations which should be stressed. However, primarily because of the algebra and differentiation involved, these topics receive only minor attention today in our courses. CAS can change this. For example, comparing and contrasting the error bound expressions associated with numerical integration (rectangular, trapezoidal, parabolic, Taylor series), or comparing rates of convergence are reasonable exercises for a calculus student having access to CAS. Emphasizing numerical integration presents a good application of polynomial approximation as well as providing students with the opportunity to use a variety of techniques to analyze error terms: symbolic computation of derivatives, graphing, and perhaps numerical approximations to estimate bounds. Using CAS as a tool to obtain derivatives, obtain Taylor polynomials, evaluate expressions, sketch graphs, etc., enables the student to become much more deeply involved with approximation questions than if he or she had to rely on hand computation.

(3) *Exercises and Test Questions*

Most exercises and test questions consist of two parts: first analyzing the question to determine what algorithm(s) to use and, secondly, carrying out the routine algorithmic manipulations. The first part, the analysis, is concerned with the understanding of concepts and is certainly the more valuable of the two parts. However, students spend far more time on the second part and thus consider the computational part to be the more important.

Using CAS to do the computation, which they can do faster, more accurately, and tirelessly, the time spent on routine homework and tests could probably be reduced to one third of the present time spent. We propose that homework exercises and test questions be divided into three categories with approximately equal weight and time being allotted to each.

(a) The traditional two part type of exercise with the computational part being done partially by CAS and partially by hand (e.g., to transform an expression to a more suitable form).

(b) Construction of examples. Students are given a set of conditions and asked to construct an example satisfying the stated conditions or show why no such example can exist. Although CAS can be used to "check out" attempts, CAS cannot give the answer in the sense that they can in a traditional type of exercise. The exercise that asks for an example requires more creative thinking on the part of the student than does the traditional exercise of type (a). Requiring students to become proficient in making up examples is an effective way to enhance conceptual understanding. For example, a student who constructs a rational function satisfying a set of asymptote and intercept requirements has demonstrated more understanding of the behavior of rational functions than if he or she had been given a rational function and asked to sketch its graph.

(c) This category includes individual and small group projects, true-false questions, proofs, outside reading, reports, etc.

It should be noted that time is *made available* for categories (b) and (c) by *turning over* most of the routine calculations to CAS.

(4) *Limitations Imposed by our Level of Algebraic Skills*

There are several ways in which our level of algebraic skills limits our understanding or treatment of calculus concepts. We consider a few examples in which the use of CAS can significantly lessen these algebraic limitations.

(a) A commonly heard lament of calculus instructors concerns the student who "can't do the algebra and thus never gets to the calculus part of the problem". A large number of students have "successfully" completed the normal precalculus sequence of courses, but have not gained the facility to carry out algebraic manipulations well enough to be successful in calculus. The resulting frustrations often cause a student to leave mathematics rather than encouraging a student to undertake the necessary remedial work in algebra. CAS can offer these students the opportunity to comprehend and work with the concepts of calculus at a meaningful level. Furthermore, their work with the calculus may provide the necessary motivation for them to remedy their algebraic deficiences.

(b) Finding or approximating the zeros of a function is often a major subproblem in analysis. The difficulty in factoring a polynomial of degree greater than two imposes a severe limitation on both the type and "size" problem that is considered. For example, rational functions that we assign to be analyzed almost never involve polynomials of degree greater than three unless they are given in factored form. We seldom assign a problem such as finding the area enclosed by the graphs of $f(x)=e^x$, $g(x)=4-x^2$, and the positive axes because it is too difficult to determine where the graphs for f and g intersect, i.e., the zero of f-g. CAS can alter this situation. The capabilities of CAS for determining zeros combined with curve sketching and the bisection algorithm, can make finding zeros one of the most widely used techniques in calculus.

(c) In numerical integration we usually "skimp" on applying Simpson's Rule because of the difficulty (in terms of time and algebra) involved in finding and bounding the fourth derivative. CAS can lesson this difficulty. Because of the algebra involved we usually do not ask students to compute very many Riemann Sums or to experiment by considering different partitions or varying the points in the partition intervals where the function is evaluated. It would be easy to carry out these types of experimentation with CAS and in the process give students a *hands on* type experience.

(d) In series, it is our difficulty in differentiating and doing the algebra that limits us to considering only a small number of *standard* functions, not the concepts themselves. For example, we would not expect a student to find the thirteenth degree Taylor polynomial for

$$f(x) = sin(e^x + 4x^3)log(cos(x) + 7)$$

although we would expect the student to know how to do it "theoretically". CAS would allow the student to actually find the Taylor polynomial and work with it.

IV. Questions and Responses

Some commonly expressed concerns about CAS and our responses follow.

(1) Question: *Will using CAS result in a serious loss of computational skills?*

Response: *No.* Using CAS will produce changes in students' computational skills. For example, there will be a definite decrease in students ability to recall and implement standard algorithms (e.g. long division, factoring, differentiation, integration, etc.). Many view this reduction as a worthwhile trade off for improved understanding of concepts. Also students' basic arithmetic-algebraic skills will improve as a result of an increased emphasis on approximation, constructing examples, analysis, and the increase in the number of exercises students work.

(2) Question: *Will freeing the student from hand computation result in a loss of confidence and sense of accomplishment?*

Response: *No.* It is true that for most (mathematically) successful students computation has been a major source of building self confidence and a sense of accomplishment. Enjoyment and success associated with computational skills have attracted many students to mathematics. It is probably also true that an equal number of students have been driven away from mathematics by their frustration with computation. We expect the ability to work problems freed from (many) present day algebraic limitations and the encouragement to explore and experiment will provide confidence and a sense of accomplishment to the first group of students and be a godsend to the second.

(3) Question: *What equipment is needed and what is the availability of hardware and software for using CAS in calculus instruction?*

Response: *There are several CAS available for time-sharing.* A small class could use one of the large systems, such as MACSYMA, REDUCE, or SMP, which were designed primarily for individual researchers and require substantial system resources. The Maple system is designed for use by medium sized classes of students on a time-sharing system. For several years muMATH has been available for microcomputers, and similar systems are being introduced. At least some venders are interested in placing CAS on hand-held microcomputers, which will allow CAS to appear directly in the classroom. Until CAS on hand-held micros become generally available, a microcomputer and projector combination can be used to facilitate the instructor's in class presentation and student faculty exploration.

V. References

1. *B.W. Char, K.O. Geddes, and G.H. Gonnet*, An Introduction to Maple: Sample Interactive Session, University of Waterloo Research Report CS-84-04, January, 1984.

2. *P. R. Halmos*, The Heart of Mathematics, American Mathematical Monthly 87(1980) 519-524.

3. *R. Pavelle, M. Rothstein, and J. Fitch*, Computer Algebra, Scientific American, 245:6 (Dec. 1981) 136-152.

4. *H.S. Wilf*, The disk with the college education, American Mathematical Monthly 89:1 (Jan. 1982) 4-8.

5. *S. Wolfram*, Computer Software in Science and Mathematics, Scientific American 251: 3 (Sept. 1984) 188-203.

6. *D.Y.Y Yun and D.R. Stoutemyer*, Symbolic Mathematical Computation, Encyclopedia of Computer Science and Technology, 15 (1980), M. Dekker, NY, 235-310.

Twenty Questions for Calculus Reformers

by Lynn Arthur Steen
St. Olaf College
Northfield, Minnesota

There is little doubt that calculus is the central subject in the mathematics curriculum. Every student with serious aspirations for a career in science, engineering, or business takes a crack at it. The mathematics curriculum of the high schools is focused on calculus as a target, and most mathematics-based courses in college use it as a common foundation. Each year about 500,000 students in the United States study calculus, about half the number who will eventually graduate from college. Calculus is big business with far reaching consequences for students, for colleges and universities, for the mathematical community, and for our nation.

The occasion of a conference on the role of calculus brings to mind a host of questions that need to be addressed. In advance of the Tulane conference, I posed 20 questions that I hoped would help stimulate discussion of important issues. At the conference few questions were answered, but more were added. So these 20 questions have now grown to 28!

1. **Should fewer students study calculus?** Perhaps society would be better served if more students were introduced instead to statistics, matrix algebra, discrete mathematics, or computer science. Does the mystique of calculus as the unique gateway to mathematics obscure the comparable value of collateral subjects?

2. **Is calculus an appropriate filter for the professions?** The

majority of students who study calculus do so en route to
careers in business, medicine, and law where calculus _per se_
will almost never be required. Is calculus really the best
means of assuring analytical skills for such students?

3. <u>Will</u> <u>computer</u> <u>science</u> <u>dethrone</u> <u>calculus</u>? For the first three
    quarters of this century, calculus has been <u>the</u> gateway to all
    significant college-level mathematics. Now that computer
    scientists urge discrete mathematics rather than calculus for
    freshman-level study, students progressing through typical
    university science and engineering curricula can no longer be
    assumed to be literate in calculus. If knowledge of calculus is
    not assumed in advanced courses, will the motivation to study it
    remain strong?

4. <u>Do</u> <u>students</u> <u>really</u> <u>learn</u> <u>the</u> <u>major</u> <u>ideas</u> <u>of</u> <u>calculus</u>?
    Physicists employ calculus as the language of science;
    philosophers talk of it as the rosetta stone of the scientific
    age; historians see it as the culmination of a millenium of
    investigation. But all students usually see (and are examined
    on) are calculations of derivatives, integrals, differential
    equations, and series. Is this because the average U.S.
    calculus student is not intellectually ready to understand the
    central ideas of calculus?

5. <u>Has</u> <u>calculus</u> <u>become</u> <u>a</u> <u>cookbook</u> <u>course</u>? In typical texts,
    exposition is subservient to template examples and carefully
    planned exercises. Routine exercises dominate tests, most being
    entirely within the capabilities of symbolic algebra programs.

Do students really learn the fundamental structure of calculus from courses dominated by calculation?

6. **Does calculus focus excessively on closed-form formulas?** Calculus in action deals with functions as they are--from laboratory instruments, from graphs, and from messy formulas. Calculus in courses deals with shadow functions--sterile, simple, artificial constructs designed to make the exercises work out well. Will students for whom the main clue to an error is a messy answer ever be able to make reliable use of mathematics in the real world?

7. **Should calculus students learn to use or to imitate computers?** Symbolic mathematics packages (MACSYMA, SMP, MAPLE, MUMATH, etc.) will soon be so widely available that algebraic work will be done as arithmetic is now--at the push of a button. How soon will this change the way in which students use their homework time in a typical calculus course?

8. **What new topics are essential for calculus in a computer age?** Asymptotic analysis, spline calculations, numerical analysis, and scientific computation have changed the way calculus is used in the scientific and engineering communities. Should the content of university calculus reflect these changes?

9. **Which topics in calculus are no longer essential?** Calculus is now over-filled, because topics are always added but rarely removed. Many topics remain only because of tradition or because they are useful for some advanced course--not

necessarily because they are crucial to the course itself. Are integration techniques any more useful than the square root algorithm?

10. <u>Do</u> <u>engineers</u> <u>still</u> <u>need</u> <u>the</u> <u>traditional</u> <u>calculus</u>? The perception, whether justified or not, that physics and engineering students must be taught the full "engineering calculus" in its traditional form impedes curricular reform now being demanded by computer scientists. Is it possible that even students in the physical sciences could benefit from a new blend of discrete and continuous mathematics in the first two years?

11. <u>Should</u> <u>calculus</u> <u>be</u> <u>a</u> <u>laboratory</u> <u>course</u>? Computer graphics now make possible a visual presentation of many of the dynamic phenomena studied in calculus. This unprecedented capability suggests wonderful pedagogical possibilities. Can we afford to provide every calculus student with access to a powerful workstation for calculus learning? Can we afford not to?

12. <u>Is</u> <u>there</u> <u>any</u> <u>reason</u> <u>to</u> <u>teach</u> <u>high</u> <u>school</u> <u>calculus</u>? About ten years ago MAA and NCTM adopted a joint statement urging high schools to refrain from offering calculus unless it can be taught at a university level--staffed by a qualified instructor, and enrolled by qualified and capable students. Despite this position paper, the majority of high school calculus courses are still thin introductions, inadequate to provide advanced placement into college courses. Shouldn't the only calculus taught in high school be university calculus?

13. <u>Why do U.S. students perform so poorly on international tests</u>?
     On the 1982 International Mathematics Assessment, only 30% of
     U.S. high school calculus students could select the correct
     answer out of five choices for the integral of a linear function
     presented in graphical form. On the whole, the top 2% of our
     students performed only at the median of the top 10% of the
     students in other countries. On the Graduate Record
     Examination, which is largely based on advanced calculus,
     foreign-educated students average one standard deviation higher
     than U.S. educated students. Are U.S. students less able, or
     less well prepared?

14. <u>Is there any value to precalculus remedial programs</u>? Much of
     the effort in remedial work is devoted to repeating early parts
     of algebra that are needed for the study of calculus, a fantasy
     that is totally beyond the grasp of most students in these
     courses. Would exploratory data analysis or elementary
     programming be better for these students, even though it
     unhitches them from the calculus bandwagon?

15. <u>Why do calculus books weigh so much</u>? The economics of
     publishing compels authors of calculus textbooks to add every
     topic that anyone might want so that no one can reject the book
     just because some particular item is omitted. The result is an
     encyclopaedic compendium of techniques, examples, exercises and
     problems that more resembles an overgrown workbook than an
     intellectually stimulating introduction to a magnificent
     subject. Would the health of calculus be improved if it were

put on a diet?

16. <u>Can one design a good calculus course from a survey</u>? Publishers now are fond of surveying calculus instructors to determine the precise blend of topics for their next calculus book. Responses to these surveys never represent a random selection, and are rarely informed by active users of calculus. Can this process possibly produce progress, or is it doomed to reflect the <u>status quo</u>?

17. <u>Is calculus a good course to train the mind</u>? Many students take calculus not for its content but for its reputation of rigor: for many it is a modern equivalent of Latin or Greek--a means to train the mind. Professional schools repeatedly use calculus as a filter, to identify students who have the right stuff. Are these perceptions warranted by the results?

18. <u>Can calculus courses convey cultural literacy</u>? Calculus is one of the great intellectual achievements of mankind, with a distinguished history of theory and applications. Few texts and few teachers can communicate much of that cultural impact in a course crammed with techniques and theory. Do courses that fail to impart the cultural significance of calculus do justice to the aims of liberal education?

19. <u>Does calculus contribute to scientific literacy</u>? Since for the majority of educated citizens calculus is the climax of their mathematical studies, one might hope that it would leave them with a good appreciation of the role mathematics plays in

society. Is calculus really the best college mathematics course to prepare educated citizens to function thoughtfully in the world of the 21st century?

20. <u>What will calculus be like in the year 2000</u>? The topical outline of calculus courses today is little changed from the early textbooks around 1700. Stability of fundamental ideas speaks to the enduring value of the subject. But the world for which we are preparing our students is profoundly different than it was three centuries ago. Will calculus be able to adapt to these differences?

These were the original 20 questions, posed in advance to those who attended the conference. The ensuing discussion of the many papers produced more questions than answers, and led me to posit these additional queries:

21. <u>Do students ever read their calculus books</u>? The shape of university calculus is largely determined by massive mainstream calculus compendiums which students are required to purchase by virtue of university-wide adoption. But do students actually read these books? Don't most students learn instead by doing problems and by discussing mathematics with other students? Would these gargantuan books sell if they were not required?

22. <u>Should precalculus be a prerequisite for calculus</u>? Too often students enter calculus without having completed a thorough four year preparatory course of high school mathematics. The result is frustration, failure, and wasted time for both student and

teacher. Shouldn't placement standards be enforced so that only those who are proficient in precalculus would enroll in calculus?

23. **Is teaching calculus most like teaching a foreign language**? It used to be assumed by administrators and teachers alike that mathematics instruction required the same intensive one-on-one interaction that takes place in foreign language classes. Isn't this still true? Wouldn't more learning take place if calculus were taught like beginning intensive French?

24. **Should the student-faculty ratio for calculus be limited**? Calculus is taught in many different class structures, from 10 students classes to 500-student lectures. There are examples of success and examples of failure at virtually all class sizes. But is there any alternative to regular, detailed feedback on student problem-solving efforts? Is there a minimum instructional effort, however it is packaged, that is necessary to insure success in teaching calculus?

25. **Do student evaluations favor calculation-based courses**? The decline in student interest in mathematics roughly parallels the increasing standardization of calculus texts, course structures, and student evaluations—all in the direction of mimicry mathematics. Is it possible that student evaluations have become an evolutionary force favoring calculation-dominated courses?

26. **Are there enough qualified calculus teachers**? To maintain

quality of content and excitement of purpose, calculus teachers should be active users of the subject. But to insure good teaching, they must be interested in their students and dedicated to good pedagogy. Are there enough teachers who meet both standards? Are there any?

27. Who will be the calculus teachers in the year 2000? Increasing demand for mathematically trained persons coupled with a rising college population in the late 1990's will push demand for calculus to record levels. But retirements from post-war faculty will also be high, and supply of new faculty--recruited from the presently depleted classes of high school and college mathematics majors--will be very low. Will calculus teaching be left to those without advanced training in mathematics--or perhaps to computers?

28. Should calculus be taught only by experienced teachers? Consistent reports from different institutions suggest that calculus too often suffers from poor teaching: inexperienced and often inarticulate teachers, excessive failure rates, disillusioned client departments, counterproductive departmental policies all point to a climate of neglect in which the most teaching is done by those with the least experience. For the good of mathematics, and for the good of the nation, shouldn't we make sure that only the very best teachers teach calculus?

What's all the Fuss about?

by

S. K. Stein, University of California at Davis

The proposal to hold this conference says that, "the teaching of calculus is in a state of disarray and near crisis... [with a] failure rate of nearly half at many colleges and universities." An alarm was sounded earlier by the January, 1985 AMS/MAA joint panel," Calculus instruction, crucial but ailing" [1].

This came as a surprise to me. Why is the teaching only of calculus under scrutiny? Are we doing such a wonderful job with discrete mathematics, linear algebra, differential equations, complex variables, or upper division algebra? Perplexed, I asked some of my colleagues, good mathematicians and fine teachers all, "What's your impression of the teaching of calculus, here and elsewhere?" One professor suggested that we might drop a couple of topics, maybe some integration techniques. Another said, we should meet five times a week instead of four but he doesn't want to. Finding no sense of calamity, I talked to colleagues in the physics and engineering departments. They liked what we do, but urged us to do more of it in the first quarter, especially differentials, vectors, $e^{i\theta}$, Stokes' theorem, and certain differential equations.

Then I went to the placement office, which helps undergraduates obtain summer internships and seniors get jobs. "What have you heard about calculus?" They were not aware that calculus is in disarray and ailing. I asked what employers were looking for. The answer was clear, "Students who can communicate orally and in writing, think, are not afraid of numbers, with a little touch of the computer." Still no complaint about calculus.

I asked my engineer son-in-law what he looks for when he recruits. His answer: "People who can deal with questions on their own." He seeks recommendations from a professor who regularly assigns his class a few open-ended problems. Though not hard in the sense that their solutions requires the insight of a genius, they are not directly related to the day's lesson. One year not one of the professor's three hundred students could solve his problems. My son-in-law did not blame calculus for this tragedy, though it was clear to me that we do little to prevent it.

So I picked up a calculus market research report that McGraw-Hill had done in 1981, based on a questionnaire sent to mathematics professors in over 200 colleges and universities of all sizes. According to the poll, 83 percent of the students in first semester calculus complete the three-semester sequence. That was reassuring. Furthermore, if there was a feeling that something was wrong it should show up in the respondents' comments on the texts they were using.   But of the 227 replies 170 judged their text's completeness to be "good" or "excellent" and only 47 called it "poor" or "adequate". They seemed quite satisfied with "topic sequence as well" with 173 out of 227 calling it "good" or "excellent."

In spite of these calming numbers, I still felt that there is indeed something in disarray in calculus teaching, something ailing. Whatever it is, we can't blame the publishers. The books they offer us respond to such polls; the manuscripts are read by a panel of independent, conscientious reviewers. We get the texts we ask for. The problem lies with us. Mathematics, the only discipline where all the cards can be laid on the table, and which therefore should be the best taught, is often among the worst taught subjects. One reason is that we haven't decided what we are teaching.

This uncertainty is visible in the discussion, The Introductory Mathematics Curriculum, presented in [1]. There we find such statements as, "We must instead teach how to create mathematics" (R.W. Hamming, p. 388), "Even more essential is the creation of courses that focus on concepts. Ideas and problem solving are the really critical part" (Robert Davis, p. 391); "Our teaching fails to provide students with the joy of using mathematics to cope with challenging problems" (Wade Ellis, p. 393); "The main fault of the introductory curriculum...is an issue of pedagogy as much as of the content" (Patrick Thompson, p. 394); "Curriculum change must be accompanied by severe questions of current teaching methods" (John Mason, p. 395). Though appearing as asides to the main debate, they call attention to what I feel is the central issue.

Before we propose the medicine, we had better agree on the diagnosis. The diagnosis depends on what we mean by "health," that is, what we are trying to accomplish in our introductory courses. That may depend to some extent on whether the course serves other majors or our own. (According to the McGraw-Hill poll, enrollment in the basic calculus runs about 60% physical science-engineering, 20 percent life science-biology-economics, 12 percent math, and 8 percent others.) In large schools the second group often has its own calculus sequence; at Davis, with its strong biological emphasis, more students enroll in the short calculus than in the engineering sequence. So the main calculus sequence we are talking about serves simultaneously engineers, physicists, computer scientists, and math majors. That is a boundary condition that any solution must satisfy. But it is not as restrictive as it may appear, since there seems to be a consensus that the students in these varied majors should learn to write, read, and think. The dean of computer scientists, E. W. Dijkstra, has written that the most important requirement for a computer scientist is mastery of his native tongue. And my computer-science colleagues urge us to expect well-written answers and proofs in our sophomore course on sets, relations, functions, and induction.

But what about calculus, where the texts have settled into a fairly uniform table of contents? There are always a few sections that the instructor may delete, such as Kepler's laws or Lagrange multipliers. But the instructor could consider deleting some more topics, such as some formal integration techniques or even related rates. Authors have less choice, for if they omit someone's favorite topic, their books will not be adopted and soon will be out of print. After all, calculus committees meet in order to reject books, much in the same way that canneries sort tomatoes. Labelling a section "optional" will surely offend someone who feels his students will then not treat it seriously if he covers it. It seems that a calculus author has the freedom to make only two decisions: Where to put analytic geometry and whether the title should be Calculus with Analytic Geometry or Calculus and Analytic Geometry. Thus the major revolution in calculus texts in the last decade has been the introduction of a second color. (In high school texts, the number of colors has reached four.) Whatever proposals this conference may make, I predict calculus will begin with functions, limits, derivatives, extrema, integrals, the fundamental theorem, go on to more applications, series, and then reach at least partial derivatives and multiple integrals. Still, there are options, and perhaps this conference will encourage publishers and professors to be more flexible when developing a table of contents or a course syllabus.

The fundamental question is not, "Should discrete mathematics precede calculus, follow it, be woven into it, or be separate and simultaneous." The question should be, "What are we trying to do in calculus and discrete mathematic courses other than cover some definitions, facts, and algorithms?" If the answer is "nothing", then we make no basic changes. If we also want the student to learn to "think" (this is now called 'problem solving' and 'heuristics') and to write, then we should act accordingly. The last thing we should do is ask for texts that mix discrete mathematics and calculus, for invariably, when two subjects are put between the covers of one book either the book grows unacceptably large or one of the two is sacrificed to the other, or both are shortchanged. Witness the fate of analytic geometry in our calculus books or of both algebra and its applications in our applied algebra books.

My own proposals may appear mild. Indeed, the first one is, but the second could encourage a change in emphasis.

The first is specific, and concerns calculus and discrete mathematics. I suggest that a discrete course of a quarter or semester be available to freshman (if that is successful, then later it could be extended). It could be taken simultaneously with beginning calculus, or alone, or, in the case of non-engineering students, with the calculus delayed. Such a course could help develop maturity and thus prepare students for calculus. It could, incidentally, weed out those who are not ready to go on. (All campuses of the University of California already require passing an exam on high school algebra and trig for entry to calculus.) It would also broaden the student's mathematical perspective earlier.

My second suggestion applies to our curriculum in general and is a response to what I see as the disarray and the ailment. Implementing this suggestion does not require new courses, nor radically new texts. However, if enough of us act on this suggestion, we may provide the quorum to support certain changes in the texts.

It too is modest, for I find that proposals for abrupt major reform tend to be carried out in form but not in substance, or viewed as something for someone else to implement.

My suggestion is rooted in my definitions of the words "curriculum" and "syllabus." Usually, "curriculum" describes the courses offered and "syllabus" lists the topics in a course. Both "curriculum" and "syllabus" call attention to the material treated. They do not refer to the way it is treated and certainly they do not mention what should be our main goal: to develop the student's ability to read, analyze, write,and speak. We easily lose sight of this objective, for facts tend to displace process. We see this bias both in the classroom and in texts. I hope that the reform suggested by this conference gives process at least equal billing with content. And I hope that authors maintain a similar perspective as they try to implement our recommendations.

My suggestion is only a modest step toward rescueing process from subservience to content.

I propose that in whatever course we teach we include a significant number of what might be called "open-ended" or "exploratory problems." Though not routine, they should not be difficult in the sense of a Putnam problem. I mean that when a student sees the solution, he will say "I should have gotten it." These problems should encourage experimentation and independent work. The answer should require the student to write coherent sentences. That means that the instructor or some other qualified person should read and evaluate what is turned in. He should demand suitable revision. The solution should not be in the solutions manual; it should not be closely tied to the particular section in the book that is being covered in class. The assignment should not be due the next day, so that the student will have time to mull it over.

Some examples will bring this proposal down to earth. To demonstrate my neutrality on the relative merits of calculus and discrete mathematics, I will choose some examples from both disciplines. I begin with examples that parallel the standard calculus.

Example 1. Let $f(x) = ax^2 + b$ be a polynomial of degree 2. Is there a polynomial g of degree 3 such that the two compositions, fog and gof, are equal?

Remarks. If the students have trouble, then you might suggest that they look at a specific f(x). Little in their earlier education has suggested such a bold step. The computations involve nothing more than cubing a quadratic or squaring a cubic. The algebra is not mysterious and the final result is both elegant and surprising. Moreover, the student should be urged to write the solution with more than a string of equations. We have a right to expect an introduction and a conclusion. We should demand that a sentence begins with capital letter and ends with a period. The left margin should be straighter than the right margin. The student may complain that such request are inappropriate in a math course. But that same student may one day be writing software manuals and internal memoranda. For us to demand less is to shortchange our students.

Example 2. Are there continuous functions f such that $f(x+y) = f(x) + f(y)$ for all real numbers x and y?

Remark. The student may or may not come up with some examples. You may have to steer him out of a rut. If he finds f(x) = kx, you might then ask, "Are there more?" (In a discrete course, the domain could be Z instead of R.) Of course one could also ask for solutions of $f(xy) = f(x) f(y)$.

Such exercises are usually delayed until the junior year, but they are appropriate during the lower division courses as well. Perhaps we could delete a few topics from the standard curriculum, whether calculus or discrete mathematics, lowering the pressure so students would have more time for this type of problem.

Example 3. Let R be a bounded plane convex set. Is there a chord that bisects its area?

Remarks. For us this is a trivial exercise in the intermediate value theorem, but most students will need help. They cannot turn back a couple of pages for the example that's just like this exercise. After this problem is solved one might ask whether there is a chord that bisects the area and the perimeter at the same time.

Example 4. What happens to x^y when x and y are near 0 but positive?

Example 5. Which polynomials of degree at most 3 have inflection points?

Remark. Much is lost in a more conventional wording, such as, "Show that every polynomial of degree 3 has an inflection point." One might then ask about polynomials of degree 5.

Example 6. Let f be an increasing positive function on the interval [0,1]. What, if anything, can we say about the centroid of the region R under the graph of f and above [0,1]?

Remark. A variant is to demand that f also be differentiable and concave down and ask about the centroid of its graph. Or we could ask whether there is any relation between the centroid of R and the centroid of the solid of revolution obtained by revolving R around the x axis.

Example 7. Let R be a bounded plane convex set and P_o a point in R. Assume that each chord of R through P_o has length at most a. What can be said about the area of R?

Remark. This question ultimately takes the student back to the formula for area in polar coordinates and extrema problems. For a discussion of this example see [2].

Now for some illustrations in discrete mathematics.

Example 8. You could compute x^6 with five multiplications by writing $x^6 = x(x(x(x(xx))))$. But you could also write $x^6 = (x^2 x^2)x^2$, which requires only three distinct multiplications. (Assume that once a multiplication is done, the result remains available.) Investigate the smallest number of multiplications needed to compute x^n.

Remark. The exact formula is not known, though eventually students can show, with the aid of an induction, that the number is at least $\log_2 n$ and equals $\log_2 n$ when n is a power of 2.

Example 9. In which linear graphs can we find a path that passes through each edge exactly once?

Remark. This is usually given in the "theorem and proof" form, but I think it far more instructive for the students to discover the result themselves. When I have raised the question in a liberal arts class, it isn't long before students observe that the vertices of odd degree give trouble and find the necessary condition quickly. Of course, sufficiency is harder to demonstrate.

Example 10. Let f be a permutation or a finite set. Is there necessarily a positive integer k such that f^k is the identity function of that set?

Remark. The approach may depend on whether this is given before or after the cycle decomposition of a permutation. In the first case the student will be more likely to experiment. That means choosing some specific sets and functions, again a traumatic experience for students not used to such freedom and responsibility.

Example 11. In a finite graph is there anything that one can say about the number of vertices of even degree or about the number of vertices of odd degree?

Remark. This exercise usually appears as a theorem. Too often we ask a question and then answer it before the student has had a chance to live with the question. By answering our own questions we turn the students into spectators, putting a barrier between them and the material. The temptation to do this is usually irresistable and is often justified by the "need to cover the syllabus." But what if the syllabus includes "teach students how to explore, to make conjectures, to write clearly"?

Example 12. Is there any relation between the number of vertices and the number of edges in a finite tree?

Remark. The comments on Example 11 apply to this example as well. In both cases we can ask the students to prove their conjectures. There are several ways to justify both, including induction. These therefore serve as legitimate induction problems. The sooner we reduce the number of traditional induction problems like, "Show by induction that $1^2+2^2+ \ldots n^2 = n(n+1)(2n+1)/6$", the better. In a realistic induction problem, the student should propose the statement to be proved. (Recall Example 8.)

The next exercise gives students far more trouble than might be expected, both in carrying out their experiments and in explaining their conclusions.

13. The function of $f: A \to B$ induces functions $F: P(A) \to P(B)$ and $G: P(B) \to P(A)$. For which f is

(a) F one-to-one?

(b) F onto?

(c) G one-to-one?

(d) G onto?

More examples discussed from a slightly different perspective are to be found in [2], but it is not hard to make up your own. Some can be derived from the statements of theorems. In some only an exploration and a conjecture are to be expected. In some a complete argument would be in order.

It may be easier to offer individual guidance and feedback in a smaller class than in a large one, but the organizational challenge in a large class should be negotiable. Though we might prefer to think our task done when we give a clear lecture, we may have to acknowledge that giving good feedback is equally important. Grading homework and examinations, which usually just offers the student the guidance of a number, is hardly adequate feedback. I suspect we, charmed by the clarity of our lectures, could go through an entire semester and never see a single page of a student's work. (I confess that this has happened with me.) It therefore may be necessary to give some time to see what the students write. It may be advisable to sacrifice content to achieve other goals.

My proposal is simply an attempt to respond to the concerns expressed by Hamming, Davis, Ellis, Thompson, and Mason that I cited. I want us to consider the goals of our teaching. Do they go beyond transmitting content? If not, we should say so in our catalogs and encourage others to introduce "problem-solving" courses to compensate for the narrowness of our mission.

If we want our students to be able to think on their own and to express their thoughts, we should give them a chance, even in the introductory curriculum, whether calculus or discrete mathematics, even in service courses, even if we propose only two or three open-ended problems in a semester. If enough of us urge publishers to include an ample supply of such problems, with variations and solutions discussed only in the instructor's manual, they will comply. But we don't need to wait for them.

References

1. The introductory mathematics curriculum: misleading, outdated, and unfair, College Mathematics Journal, Vol. 15, November 1984, 383-399.

2. S.K. Stein, Routine Problems, ibid, Vol. 16, November 1985, 383-385.

PHYSICAL SCIENCE AND INTRODUCTORY CALCULUS

James R. Stevenson

Executive Assistant to the President

Georgia Institute of Technology

Atlanta, Georgia 30332

I. Introduction

In speaking for the physical science community I need to preface this position paper with several background comments. As an experimental physicist I have a definite bias toward the intuitive understanding of mathematics as applied to the physical sciences. During my thirty years at Georgia Tech in both professorial and administrative roles I have probably been sheltered from some of the seeming problems facing the teaching of introductory calculus at other institutions. This sheltered existence could possibly be of use in the current workshop.

Early in my career as a faculty member and as an academic administrator, I learned quickly that dictating course content or teaching approaches to faculty within physics was a uselesss waste of time and if you tried to impose standards on other disciplines you quickly learned how fragile "Humpty -Dumpty"is!

This workshop and this paper will deal in a lot of rhetoric never
to find the light of general acceptance, but if reasonable
concerns can be delineated the possibility of multiple solutions
will exist to be formatted to individual situations.

II. Impact of Department Evolution on Service Courses.

 Let me start with some observations from my experience in the
discipline of physics which I think has a rather close
correspondence to the apparent problems in introductory
mathematics instruction. In the instance that a physics
department at an institution is without a degree program or has a
degree program with a limited number of majors, the department
tends to be dependent on its service role. The faculty is
usually small and much of the faculty dialogue is across
discipline lines. The number of majors expands, a graduate
program is added, faculty become more involved in research. The
faculty dialogue becomes more in-grown and the concerns are
shifted from the service role to preparing physics majors for
graduate school. The content of the introductory courses is
changed. Items such as fluids, kinetic theory of matter,
statics, introductory thermodynamics and other such items are
eliminated to provide the proper background for the orderly
progression of the physics major through the discipline and to an

adequate preparation for graduate school. After a number of iterations at Georgia Tech, we found that we had an excellent undergraduate program for preparation of physics majors for graduate school, but we had a significant number of majors going directly into industrial positions with a bachelor's degree. As we did not create separate courses for our service load, many students received a rather sterile picture of an exciting discipline. A few attempts were made to accommodate the majors interested in a terminal B.S. degree within the existing program. Any major modification was not possible and the only alternative was to construct a second degree program. The birth of a program in applied physics resulted but no changes were made in the introductory courses. A number of disciplines feel that we could be of better service to the non-physics majors by offering a separate course sequence designed for engineers and other science disciplines. Having failed in convincing the majority of physics faulty we now find a number of disciplines employing physicists and teaching the elements of quantum mechanics and other such topics. I think you will find similar evolutions of other disciplines including mathematics. Is this type of evolution and end result good or bad? The eyes of the beholder must make the comparison and the decision.

III. The Role of Empiricism in Faculty Attitude

In the situation previously described, the faculty are
relatively happy with teaching the introductory courses as they
understand the objectives and have controlled their
implementation. The courses are frequently revised and some
thought is given to more effective pedagogy. Physics has a
significant difference from mathematics in that it is an
empirical science and can not stray too far afield from
observations; thus the connectivity to engineering and the other
sciences is always present. More empiricism is becoming present
in mathematics with the computer being used as an experimental
tool to gain insight into solutions of non-linear equations. The
title "applied mathematician" is accepted by many and scorned by
others. Whereas the theoretical physicist is dependent on the
observations of the experimentalist, the "pure" mathematician ha
no such dependence. A schism can develop between faculty members
and honest differences of opinion can have impact on the
introductory curriculum.

Another factor impacting faculty attitude in many
institutions has been the assessment criteria for adequate
credentials within institutions of higher education. We have
gone to the opposite extreme of the secondary schools in which

the motivation to teach and the knowledge of how to teach replaces scholarship in the subject matter to the college-university extreme that a PhD in subject matter and demonstrated scholarship in the field transcends the ability or motivation for effective communication in the classroom. Frequently one observes a misplaced scholar in a junior college environment unable or unwilling to effectively communicate in the environment of an introductory course although the scholar might contribute in an industrial or graduate school environment. The need for credentials and accountability of performance need to be reconciled.

As mentioned earlier the rhetoric of this workshop will not result in changes in the existing system nor in the attitudes of current faculty. The need is to understand the present status and suggest alternative paths rather than carry on an evangelic revival.

My premise is that departments of mathematics dominated by "applied mathematics" are not having the same reaction to the apparent inadequacy of introductory calculus instruction as those departments dominated by "pure" mathematics. The presence of the geometrical interpretation of calculus with frequent reference to applications does remove sterility from the subject as taught as a service to other disciplines. Many faculties do not have sufficient depth to provide this type of insight at the introductory level and even less interest in developing pedagogical materials. In addition some departments of mathematics have a "critical mass"of faculty for scholarship research into some area of pure mathematics and a distortion of this "critical mass" for service teaching is not necessarily wise.

IV. Content and the Intuitive Approach.

Of equal importance to content in the introductory calculus is the geometrical and intuitive feeling for mathematics. In an article (1) published in 1971 entitled "Mathematics for Physicists: A Report on the National Study of Mathematics Requirements for Scientists and Engineers" by G.H. Miller, a survey of 929 physicists indicated that a great majority prefer a mathematics course which is approximately 50% theory and 50% applications. Now applications in themselves can be handled without spending time in developing an intuitive feeling for calculus. However the application of calculus to the empirical sciences implies the finite limitations imposed by the act of measurement and hence never achieving the infintesimal limits of the purist. E. Leonard Jossem (2) in a 1964 paper entitled "Undergraduate Curricula in Physics: A Report on the Princeton Conference on Curriculum S" makes the following statement.

"The other aspect concerns the problem of
obtaining an optimum balance among the
various elements which must go into an
intellectually vigorous program; in
particular, the balance between the
elements of synthesis and insight into
physical situations on the one hand and
sophisticated mathematical analysis on the
other. The dangers of serious imbalance were
pointed out in connection with curriculum R.
In the experience of many of the conferees,
R curricula in which the main emphasis has
been placed on very formal, detailed, mathematically rigorous analysis tend to stullify
the imagination and to produce a degree of
intellectual rigor mortis even in very good
students."

The physical scientist needs the facility to handle sophisticated
mathematical analysis and abstract thought but with a constant
realization that the correctness of the mathematics does not
imply the correctness of the physical model to which the
mathematics has been applied.

The meaning of slope and curvature and their relation to the first and second derivative are important. The location of maxima, minima and inflection points are also important, and the physical scientist must have sufficient drill to be able to look at a graph and tell immediately the sign of the first and second derivatives as well as estimate the magnitudes without resorting to computers. In a similar vein the area under a curve must have an intuitive relation to integration. Infinite series are frequently used to approximate analytical functions. Some knowledge of convergence as well as truncation errors are needed on an intuitive basis prior to releasing the power of a computer to grind away and produce nonsense.

Every faculty member in a discipline served by mathematics will have a unique perspective as to the essential content of the introductory calculus sequence. In 1963 the recommendations of the Second Ann Arbor Conference (3) for physics majors includes:

> "Vector analysis,including gradient,divergence,
> curl, Laplacian, together with their physical
> significance. Line and surface integrals, Gauss
> and Stokes theorems. Vectors in Cartesian,
> cylindrical, and spherical polar coordinates.
> Some knowledge of existence of other orthogonal
> systems and of physical applications of matrices
> and tensors."

A year later the report (4) of the Princeton Conference on a more general curriculum, the phrases "Develop vectors and calculus as needed" and "Develop accessory math as needed" are used. Rather infrequently do physical scientists encounter Fourier series and Fourier transforms in mathematics prior to their use in course work within their discipline. The topic is not unique and as another example very few universities have all of their statistics courses taught by the mathematics faculty. The content of the introductory calculus courses should not be determined by faculty outside mathematics. The development of mathematical maturity, satisfaction of prerequisite topics for more advanced mathematics, and development of intuitive or geometrical understanding are the important factors.

What about the role of the computer in the introductory calculus? Many of my colleagues feel that the utilization of the computer needs to be very carefully controlled in the introductory courses. Physical science departments frequently use computer simulations to replace laboratory experience. This use to the exclusion of laboratory measurement is far more dangerous to the scientist than anything the mathematician could do in the introductory calculus sequence. In the same vein the use of computer simulations in the introductory calculus sequence with the intent to provide the intuitive insight to mathematics should be avoided. A computer laboratory experience in which the student

is first confronted with a graph to analyze by hand followed by
computer analysis would be an appropriate use which would
reinforce the intuitive approach and give understanding to the
value of the computer in non-analytical situations.

Computer assisted instruction does require a few comments. As
with most service courses, introductory calculus is not going to
cover the specific content requirements of all disciplines
dependent on the calculus sequence. Computer assisted
instruction can provide a valuable adjunct to the textbook for
modular self paced instruction of students with a requirement to
learn a mathematical technique or concept outside the classroom.
For example the curl or divergence of a vector quantity may not
be covered adequately for a student in physical science or
engineering. The development of a library of modular topics to
enrich and broaden the content of introductory calculus is
desirable and could be available in a central library, branch
library, mathematics laboratory, learning center or combination
of the above. The introductory sequence must provide the
mathematical maturity and confidence to insure a reasonable
degree of success by students needing to supplement the content
of the introductory course. Students in the introductory
sequence should be assigned a topic from the modular material to
monitor the system, provide feedback, and to give added
confidence.

V.   Introductory Calculus-Quo Vadis?

Professor Douglas has set the tone for this workshop in a very simple and pragmatic manner.   In a letter he states, "My aim in running this conference is practical, that is, I'm less interested in 'what is best' than I am in 'what is possible'." One can contend that the best is always possible, but the meaning and context are obvious.   From the point of view of physical science I would like to address four different but interconnected topics addressing a possible approach to the perceived need for more rapid evolution of introductory calculus education.

1. Extended Faculty.

Assuming that there is some value to the influence of "applied mathematicians" to the content development and instructional philosophy in introductory calculus, a method needs to be devised to provide this influence without distortion of the existing mathematics faculties and in the face of shortages caused by the demands of industry and other related disciplines.

In urban environments the selective use of industrial mathematicians as instructors in the introductory calculus can be helpful; however, part time instructors are band-aids and do not have the continuity or motivation to address the illness. A more

effective approach is to develop a cadre of "intoductory applied

mathematicians" from other disciplines on the campus such as

engineering and physical sciences. This cadre should have part

of their time assigned to the mathematics department to work with

selected faculty in mathematics in both teaching and the

development of curriculum materials. The assignment must be

mutually acceptable at all levels and should be of a finite

length with clearly described objectives so as not to threaten

existing boundaries and reward structures. The cross breeding is

already present in many disciplines. Chemists and biologists are

present in physics departments, mathematicians and physicists are

found in electrical engineering,nuclear engineering, and

mechanical engineering. Many disciplines are found in

departments of computer engineering and computer science.

Mathematics has remained relatively pure.

2. Content and Pedagogical Approach

As might be expected from the prior discourse, a concern is

expressed not so much for the detailed content of the

introductory calculus as for demonstrating the excitement of

applications and the geometrical interpretation of the calculus.

Two quotations are of some relavance. Prof. Jossem (2) refers to

a quotation from an address by J. C. Maxwell (5) to the British

Association in 1870:

"For the sake of persons of these different
types, scientific truth should be presented
in different forms, and should be regarded as
equally scientific, whether it appears in the
robust form and the vivid coloring of a physical
illustration, or in the tenuity and paleness of
a symbolical expression."

The physical scientist would argue that a similar statement can
be made for presenting introductory calculus. Note that both the
pure mathematics approach and the applications supplement are
necessary. A danger does exist in stressing applications and
intuitive understanding to the exclusion of the beauty of
mathematical logic. In a recent interview reported in the
August, 1985 issue of "Optical Engineering Reports", (6) Dr. H.
John Caulfield, a well known optical scientist responded to the
following question: "What were the basic understandings that the
holography community had that you didn't that made it easy for
them to understand holography at that time?" Dr. Caulfield's
response was:

"In the early holography days it was the
Fourier transform analysis of all the work.
Strangely enough, no one told us that when
we were doing quantum mechanics, we were
really doing Fourier analysis. So, I had to
go back and relearn a lot of mathematics. Now

it turns out that the mathematics could quite

possibly have been stated in other ways, but

it was stated the way it was because holography

grew out of, in turn, the interferometry club

of years before, as well as the earlier

electronics signal processing community.-----"

Thus mathematical techniques taught by physical scientists for a

particular application have the danger of not being recognized

for the breadth of their applicability. The content of the

introductory calculus must be the decision of the mathematics

faculty as they recognize the logical development of the

discipline. Honest dialogue with other disciplines including the

physical scientists should be part of the decision process.

Regarding textbooks several observations can be made. First,

textbooks invariably have too much material and the question

becomes which material should be eliminated. The best approach

is to seek unanimous approval of topics to be included. The

extended faculty should be used in this decision process.

Secondly, the best introductory texts are frequently written by

faculty with more interest in teaching than in research. The

Feynman lectures and the Berkeley series are not widely used

introductory physics texts. Although the publishing community

has resisted modular textbook material, a greater pressure is

developing for this approach and much recommends modular material

as a way to initiate textbook writing as well as enrichment

material for classes.

3. Computation

Introductory calculus should not be turned into a course on
computer programming but the use of the computer as a tool for
computation of problems not lending themselves to solutions in
closed form should be demonstrated. Care needs to exercised that
the student not lose sight of the forest for the trees. The area
of computer assisted instruction for both drill and self paced
modular instruction is encouraged.

4. Dialogue

The importance of continuing dialogue between the mathematics
faculty must be stressed. Even though very few suggestions from
outside are adopted, the exercise is worth the effort. Many
times the ordering of the content of the introductory calculus
can relieve faculty in other disciplines of having to introduce
mathematical concepts. The result can be an educational delivery
system which is much more efficient.

(1) G.H. Miller, Amer.J.Phys.39,1006 (1971).

(2) E.L.Jossem, Amer. J. Phys.32,491 (1964).

(3) Committee of Second Ann Arbor Conference,Amer. J. Phys.30,339
(1962).

(4) "Scientific Papers of J.C.Maxwell"(Cambridge University
Press, Cambridge, England,1890),Vol. 2, p. 220.

(5) R. Feinberg,Optical Engineering Reports,No. 20, August 1985.

POSITION PAPER

CONFERENCE TO DEVELOP ALTERNATE CURRICULA
AND
TEACHING METHODS FOR CALCULUS
AT THE
COLLEGE LEVEL

TULANE UNIVERSITY, NEW ORLEANS
January 2 - 6, 1986

METHODS AND STRUCTURE OF TEACHING CALCULUS EFFECTIVELY

Presented by:
Steven S. Terry
Ricks College, Rexburg, Idaho
January, 1986

METHODS AND STRUCTURE OF TEACHING CALCULUS EFFECTIVELY

INFORMATION ABOUT TWO-YEAR COLLEGES

To increase the readers awareness of two-year colleges and the challenges facing them some pertinent information will be provided the reader. The information was provided by A. David Allen of the American Physics Teachers Association.

1) Two-year colleges go by three different names:
   Community Colleges
   Junior Colleges
   Technical Colleges

2) There are about 1,221 of them. The total enrollment is about 9 million.

3) They employ about 250,000 teachers of which 143,000 are part-timers.

4) Most were started in the 1950's and 60's.

5) They cater most offen to:
   first generation students,
   minorities,
   lower income,
   students with lower high school grades,
   females (52% of the enrollments),
   older students (average age is 28),

6) Of all adult Americans, 8% will take a course in a two-year college this year.

7) Of all students enrolled in American higher education, 33% are enrolled in a two-year college.

8) Of all minorities that are enrolled in American higher education, 65% of them are enrolled in a two-year college.

9) Most two-year college students have aspirations for a BS or a BA degree.

10) Two-year colleges have major problems with retention and a high attrition rate.

11) In the 1980's, the faculty of two-year colleges is:
    aging rapidly (average age is increasing),
    generally lacking industrial expereince,
    and are teaching oriented.

12) Occupational area degrees represent 62% of all degrees conferred.

196

13) Articulation and transfer of courses to four-year
institutions is a major problem.

14) Most two-year colleges do not have exit or competency
exams.

15) Most students require remediation in:
science (30%),
math (35%),
English (28%).

16) Many cater to rural students.

17) The size of two-year colleges shows a wide diversity:
Idaho has only 2,
Virginia has 21,
while California has 105.

18) Full-time enrollments vary from a low of 125 to about
22,000.

19) Fifteen percent of the faculties hold doctorates while
in chemistry 40% have reached this level.

20) Fifty-five percent of all entering students who begin
college, start at a two-year college.

21) Of all baccalaureate degrees awarded in California in
the 1982-83 academic year:
21% of the university system graduates (21,328)
had attended a two-year college.
50% of the state system graduates (42,959) had
attended a two-year college.

22) Minorities account for 20% of two-year college
enrollments:
43% of all Black college students,
54% of all Hispanic students,
43% of all Asian students.

INFORMATION ABOUT RICKS COLEGE

To understand the institution where my teaching methods are
used, the following information is furnished. Ricks College is
located in southeast Idaho. The population of Rexburg is about
13,000 and the college has an enrollment of 6,500 full-time students.

The college is the largest private two-year college in the
country. It is owned and operated by The Church of Jesus-Christ of
Latter-Day Saints (Mormons). Students come to Ricks from every state

in the nation and usually about 35 foreign countries are represented
by students. The campus covers 255 acres with 46 buildings.

The college has an "open-door" policy and offers an Associate in
Arts & Letters and Associate Degree in 35 specialized disciplines.
The college has a number of one-two-and-three year technical programs
in a variety of fields.

A STATEMENT OF BASIC PHILOSOPHY

"He who has access to the fountain does not go to the water
pot." -- Leonardo Da Vinci

The method being presented in this paper is predicated on the
premise set forth in the above quotation. The underlying principle
involves the philosophy that every student needs to become
responsible for his or her own education and needs to be guided to
discover how to reach the fountain for him or herself.

Edgar Dale is quoted as saying "Life is not a hundred-yard
dash, it's a long-distance race. It's a race against sloth,
ignorance, apathy, the willingness to take things as they are and
leave them that way. But if we are to take things with gratitude
instead of for granted, we must spend some of our time comtemplating
where we are and where we are going - or drifting."

This conference/workshop presents all of us with the opportunity
to quit "drifting" and to spend a concentrated effort to contemplate
exactly where we are in calculus education and where we want it to
go.

BASIC ELEMENTS OF THE PROGRAM

PURPOSE:

To Improve The Teaching & Learning Situation in the Classroom

Every concept taught has a stated purpose so that the student
knows exactly why they need to know the information. This provides
motivation for the student to pay attention to what is to follow.

It has seemed that much of what we do in a typical calculus
classroom can be characterized as a T - T - T cycle: Teach - Test -
Terminate. The way of approaching instruction presented in this

paper attempts to change to a four cycle method of instruction: C -
E - T - E. This representes Capture - Expand - Teach - Evaluate.
This process has seemed to increase a student's commitment, attitude,
skill and knowledge of mathematics.

CENTRAL MESSAGE:

        The main concepts or principles are clearly defined so that the
student clearly knows which points are or major importance and which
have lesser value. Every student has experienced studying "the
wrong" things for a test. Usually this is because the instructor has
not made it clear just what is meat and what is chaff.

The Advantages Are:

Students Are Taught How To:

        *   Organize the material better;
            *   Retain it longer;
      *   Expand on the concepts of the class;
                *   Teach others;
        *   Evaluate what they have learned.

Teachers Tend To:

        *   Teach fewer concepts themselves;
            *   Reorganize material;
    *   Be more pleased with the students interest in the course;
    *   Find they are spending less office time helping students.

        The Central Messages for this paper are presented on the next
pages.
        1. Every Person Is Both A Teacher And A Learner.

The Three Person Problem.

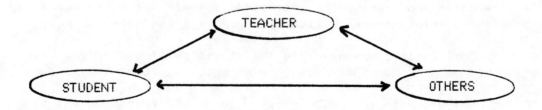

        The reason this is called a problem is because it is something
which has to be solved. How do you get the three-way interaction
between a teacher, student and others going effectively?
        This learner/teacher framework is vital to improve the four
fundamental experiences of the Four-Fold and the communication of

their essential information. This will be expanded on later in this
paper.

2.   We Improve Our Life In Four Areas.

Knowledge.

Skills.

Attitudes

Commitment.

3.   There Are Four Fundamental Experiences Associated With
     Teaching And Learning.

CAPTURE

Capture --   acquiring significant information from any
             source.

Often much of what we do in a calculus class can be expressed by
using a Charles Dickens quote. Here Dickens is describing Thomas
Gradgrind in M Chaokim Child's Schoolroom.

         "He seemed a kind of cannon loaded to the
     muzzle with facts and prepared to blow them clean
     out of the regions of childhood at one discharge.
     He seemed a galvanizing apparatus too charged with
     a grim mechanical substitute for the tender young
     imaginations that were to be stormed away."

During a Capture experience a teacher lectures in whatever style
is most comfortable for that individual. Some of the methods used by
different individuals to reinforce capture include such things as:

Having the student take notes in outline form. Sometimes the
student is required to show these notes to a group leader (steward)
to receive points and in some classes to receive a copy of the
lecture notes provided by the instructor. These notes provide a way
for the student to check the accuracy of their own note taking.

Some instructors have had the student fill out a form (see
attachments) to demonstrate that the material has been captured.

The students are usually asked to do whatever problems have
been assigned by the instructor before the next class meeting.
EXPAND

Expand -- add, integrate or apply information by yourself.

Expansion experiences are where a teacher allows a student to expand, restate, disagree, research the literature, find other examples or applications not covered in class, or whatever it takes for the student to personalize and internalize the concepts presented in the capture experience.

One instructor sets up his expansion experiences in the following way:

\#   Students might spend the first 15 minutes of the class answering questions about the homework assignment.

\#   The next 10 minutes might be devoted to asking questions of the students to test their capturing of the concepts of the previous lesson.

\#   The last 25 minutes of class might be given for students to work on some expansion experiences, either individually or in a group experience.

Some teachers have found it beneficial to use a form to record the expansions an individual might make. A copy of one used by an instructor is provided with this paper.

These expansion ideas might be from personal experiences or material from the textbook or from resource books.

TEACH OTHERS

Teach -- discuss or share the information learned with others
        to benefit both persons.

Everyone has probably heard the expression "The teacher always learns more than the student." Try to imagine how different your attitude towards your comprehension of this paper would be if I were to tell you that tomorrow you were to teach the material contained herein to someone else? Do you think you would make a greater effort to master the material? From my experience I have concluded that this is exactly what happens to a student when they are faced with having to teach the material to someone else.

Teaching experiences may be one-on-one or teaching in small groups. The groups teaching are usually informal in nature. The students help each other learn concepts or how to solve problems pertaining to the concepts. Here, both the teacher and learner benefit from the experience.

Another teaching experience is often provided by having students do problems on the board. They might explain the solution to a problem to the entire class or sometimes as many students as can be conviently accommoded at the boards are asked to work problems simultaneously. The students siting at their desks are encouraged to ask a nearby "board student" for clarification or explanation of points not clearly understood.

At first many students dread this type of exposure, but after a short period of time, most students regard this as one of the most beneficial parts of the class.

EVALUATE

Evaluate -- to improve life-long learning.

"I feel like a droplet of spray proudly poised for a moment on the crest of a wave undertaking to analyze the sea." -- Will Durrant.

Evaluation can be in traditional forms. The homework can be corrected and graded, tests can be given, notes can be awarded points and quizes can be administered.

4.   Experiences Can Be Expressed In Four Categories.

PURPOSE -- A clear statement of what the lesson will accomplish

CENTRAL MESSAGES -- A roster of the most important concepts in the lesson.

VALIDATIONS AND APPLICATIONS -- Justification and applications of the purpose and central messages.

VALUES -- A statement of how the learning of the information benefits the learner.

FOUR-FOLD SUMMARY

	Capture	Expand	Teach	Evaluate
Purpose				
Central Messages				
Valid. & Applic.				
Values				

Validations

1)   San Jose State University with Dr. Walter Gong.

Dr. Gong who was the developer of most of the concepts of the four-fold has taught his physical science classes at San Jose State University for several years with this method. His students consistently scored significantly better than two other instructors teaching in a tradition T-T-T cycle.

To demonstrate his confidence in the method he took 30 students who were identified as most likely to fail based on their previous educational experience and test scores. They were enrolled in one of his beginning biology sections. Only one student scored below a B from this group.

When it was suggested that this difference might be to the nature of his evaluation system, he had his students take the final exams prepared by the other two instructors for their classes. Gong's students did significantly better on the tests than the other instructor's students did.

2) Sabatasso Foods in Santa Ann California

The president of Sabatasso Foods explained to me his formula for the successful training of his sales staff. While hearing his technique I was struck rather forcefully by how much it resembled this four-fold process.

In the Capture phase, a prospective salesman was required to work in the bakery, processing plant, warehouse, office and then go with an experienced salesman as an observer. This was done so that the salesman would have an complete understanding of the operation of the business.

Under Expand I could easily place Mr. Sabatasso's instruction to the salesman that he must learn his own sales approach. He was to modify, change, personalize, expand, etc. the sales approach he had witnessed with the experience salesman. He had to find something which was his so that he would be able to identify with it.

The Teach part came when the salesman had to present his sales pitch to the president of the company. Here he was subjected to most, if not all, of the most common objections a prospective buyer might have for not buying the product. He was put through his paces, tested, and tried by the president before he was allowed to go out on a call.

Evaluate came into play when he would have his regularily scheduled meeting with the president. He would have to account at this time for his progress, performance and sales.

This company grew in 20 years from a one man operation (Mr. Lou Sabatasso) to one employing hundreds of people in several states and having an annual sales of about 30 million dollars a year.

3) Essentials of Education

The leaders of several professional organizations reached the conclusion that Society must reaffirm the value of a balanced education in 1978. The National Council of Teachers of Mathematics and the National Science Teachers Association were part of this group. They circulated a statement on the essentials of education among a number of professional associations whose governing boards endorsed the statement and urged that it be called to immediate attention of the entire education community, or policy makers and of the public at large.

The statement embodied the collective concern of the endorsing associations. It expressed their call for a renewed commitment to a more complete and more fulfilling education for all.

From this statement one would find, "The independence of skills and content is the central concept of the essentials of education. Skills and abilities do not grow in isolation from content. In all subjects, students develop skills in using languages and other symbol systems; they develop the ability to reason; they undergo experiences that lead to emotional and social maturity. Students master these skills and abilities about science and mathematics, history and the social sciences, the arts and other aspects of our intellectual, socal and cultural heritage.
"More specifically, the essentials of education include the ability to use language, to think, and to communicate effectively;...to acquire the capacity to meet unexpected challenges; to make informed value judgements; to recognize and to use one's full learning potential; and to prepare to go on learning for a lifetime."

It seems to this author that the elements of the Four-Fold meet the critera set-forth in this document.

4)   Brigham Young University

BYU experimented with the Four-Fold method of learning material through a forum experience. A group of freshman psychology majors were instructed in the CETE procedure and a control group of graduate physchology students were left to their own devices to capture the material presented in a once a week lecture.

At the end of the semester when the two groups were tested on the lectures, the freshman students showed a significantly higher mean score than the graduate students.
In another experiment, Grant Harrison, professor of instructional science at BYU who was trained in the CETE approach tried it on 12 first grade classrooms of the Alpine School District in the form of a pilot reading program.

Quoting from his report "The program is designed for use by the students' regular teacher and works on a companion study concept. Each student interacts with another student."

"...The children are ecstatic with the idea of being a teacher to one another. It gives them a sense of pride...Most reading programs do not have a feedback system. With this program children learn more rapidly because they are accountable to their parents."

5) Ricks College

When my students were asked to rate the course and instructor following instruction in a math class using the Four/Fold Framework, the following results were obtained:

	PRE F/F	POST F/F
Overall rating of instructor and course	2.8	3.0
Interest in course	2.5	2.8
Class helped me gain new knowledge, skills, or ability	2.6	2.9
	N = 49	N = 70
	1 Section	2 Sections

Scale:

Excellent = 3
Adequate = 2
Less than
  adequate = 1

Some Student Comments:

"Through this course I'v gained a better understanding and knowledge in the area of mathematics. Also using this four-fold method has helped alot in my understanding and remembering the concepts taught."

"The most significant concept was that grades are not the most important thing in life, but rather just learning and being able to teach others. By learning and being interested in the subject I do better in my classes."

"The four-fold has been terrific. It has been really helpful in making us expand on our lessons. I know when I do, I learn so much more. There have been times when I just wish I could sit and learn and learn about a subject. The expansion is a perfect way to do this."

"I think the most important thing I learned is that its easier and better to learn if you know the 'purpose' and 'value' of what you are learning. It creates more interest and you are not just doig things just because that the way its done."

"It has helped me by giving me a way to organize the notes of what a teacher says."

Typically Calculus classes at Ricks College experience a drop-out rate of from 10-15% which is often better than many other departments on campus.

An inservice workshop conducted in 1981.

At the end of a week long inservice workshop held for instructors from Brigham Young University, Provo; BYU - Hawaii; and Ricks College to teach this method the following comments came from the evaluation sheet given at the end of the fifth day:

"The information does not threaten the other good things I am already doing, and is adaptable to my own needs."

"The ideas have great flexibility for home, school, church and my own personal learning."

"Brought new committment, desire and a plan to carefully examine my whole performance in T/L process."

"The examples and emphasis given to expansion and teaching by students coupled with the idea of group study will help me share my purposes and values with my students."

"Our real challenge is to help all students reach 100% capture. Hopefully this workshop will help me to do this."

"The information from Gong dignifies the student and treats him as an intelligent learner, responsible for a greater role in the T/L process."

"I have realized anew the need for application."

"The three person model of Gong agrees with other carefully prepared models particularly Scouting MOL model."

"Clarity of method and presentation and the conspicuous absence of pedantic jargon was refreshing."

6) S.N. Postlethwait at Purdue University

Writing in the January 1984 edition of Engineering Education Dr. Postlethwait states that Dick Stewart of the Purdue Placement Center cites three major characteristics that employers are concerned about:
* That employees are able to take an assignment and
  get the job done
* That they are able to write a succinct report
* That they are able to make an oral presentation to the
  boss or a committee.

To accomplish this he has developed an audio-tutorial system
which employes the framework being discussed in this paper. He only
lectures for a half-hour as his belief is that content delivered
during the last half hour of a session was usually not included in a
sutdent's summary of the lecture. He cited a study where a
progressive decrease in the inclusion of content was noted as the
length of the presentation increased. We believes that with such a
steeply declining curve, probably little of the content in the last
20 minutes was absorbed by the students.

With this information his course consists of three elements:
The lecture ("broaden their perspective"), the small group session
("you really learn a subject when you teach it") and individual study
("a situation that excites student intellect").

Applications

Can be applied to all fields of teaching and learning.

Value

* Students learn more and retain it longer.

* Students become more verbal and demonstrate more confidence.

* They tend to be more helpful to peers and take an interest
  in how they are doing.

* A student tends to work harder than they would have
  otherwise.
* Students learn how to learn.

IN SUMMARY

There are fundamental experiences necessary for change or
progress in life.

Every person is both a learner and a teacher in four
fundamental areas (commitment, attitudes, skills, and knowledge)
using four fundamental experiences.

The four fundamental experiences are: Capture, Expand, Teach
and Evaluate.

All through learning includes these four activities. Understanding these leads to greater commitment and the learning curve seems to increase rather than decrease after a class is over.

The knowledge structure for organizing material is in four categories: Purpose which is motivation, questions or problems; Central Messages which are main ideas, thesis statement, the most important concepts; Validations and Applications or verification, proof, uses; and Values which demonstrate benefits, blessings, worth.

The L/T framework involving three persons will improve the four fundamental experiences and the communication of their essential information.

"Instead of taking possession of man's freedom thou did'st increase it and burden the spiritual kingdom of mankind with its sufferings forever. Thou dids't desire man's free love that he should follow thee freely enticed and taken captive by thee. In place of the rigid ancient law man must here-after with free heart decide for himself what is good and what is evil having only thy image before him as his guide. But did'st thou not know he would at last reject even thy image and thy truth if he is weighed down with the fearful burden of free choice? . . . We have corrected thy work and founded it upon miracle, mystery, and authority and man rejoiced that they were again led like sheep and that the terrible gift that had brought them such suffering was at last lifted from their hearts." -- Dostoevski from the <u>Grand Inquisitor.</u>

A 4-FOLD FRAMEWORK FOR LEARNING, TEACHING ...

TOPIC:_____   COURSE:_____
TEACHER:_____   DATE:_____

C.   TEACHING OTHERS FOR THEIR BENEFIT
Summarize your teaching experience by answering the 4-Fold questions below.

Situation: Who did you teach or discuss your ideas with?_____
            For how long?_____ Describe the situation._____
_____

1. PURPOSE. What improvements were you trying to achieve?

2. CENTRAL MESSAGE. Of all your choices, what main experiences did you choose to
   help achieve the desired improvements?

   What information did you choose to teach or discuss? Check one or more:
   Purposes____Central Message(s)____Validations/Applications__Values____
   Present in detail the most significant information you taught.

3. VALIDATIONS AND APPLICATIONS. What were the observed improvements (or lack
    of improvements) in yourself and/or other(s)?

    How could you apply any of the above information to benefit other situations?

4. VALUES. How worthwile was this "Teaching Others For Their Benefit Experience?"
    Check one: superior____ excellent____ average____ poor____ How could you
    improve its value?

A 4-FOLD FRAMEWORK FOR LEARNING/TEACHING   NAME_____
TOPIC:_____   COURSE_____
TEACHER:_____   DATE:_____

A. CAPTURING INFORMATION
Write concise and complete answers to the 4-Fold questions below.

( )1. PURPOSE: Did the teacher include his/her purposes (i.e., goals, motivations,
       needs, questions to be answered, etc.?) YES___ NO___ If yes, state them briefly.

( )2. CENTRAL MESSAGE: Did the teacher propose significant propositions, principles,
       procedures, etc., for achieving the purposes? YES___NO___ If yes, concisely
       write the central message being proposed.

( )3. VALIDATIONS AND APPLICATIONS: Were validations and applications (i.e., evidences
       authorities, examples, laws, personal experiences, etc.) cited to support or to
       not support the central message? YES___NO___ If yes, list them and explain the
       important ones.

( )4. VALUES: Did the teacher state or show nis/her/personal values (or feelings,
       choices, views, etc.) about the information presented or related issues?
       YES___NO___ If yes, summarize the values indicated.

TOPIC:_____   COURSE:_____
TEACHER:_____   DATE:_____

B. EXPANDING THE INFORMATION
Report how you expanded your learning for your own purposes and values.

1. Check one or more of the following to indicate how you expanded the captured
    4-Fold knowledge structure.
    ___a. I added new information to the 4-Fold.
    ___b. I applied the 4-Fold information to different situations and problems.
    ___c. I integrated, or synthesized, other knowledge to the 4-Fold.
    ___d. Other_____.

2. Check one or more of the parts of the 4-Fold that you expanded:
    Purposes___Central Message___Validations/Applications___ Values___

3. Write one or more significant paragraphs to describe the extent of your expansion.

More Questions for Expansion. Ask two questions you would like to know more about
                               or discuss with someone.
Question #1:_____
_____.
Question #2:_____
_____.

PRACTICING FOR LONG-TERM CAPTURE:

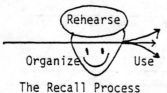

The Recall Process

1. How long did you practice the <u>recall</u> of the information
captured in the 4-Fold Framework? _____minutes.

2. How correct, complete and concise was your recall?
Check one: excellent___ average___ poor___ no practice___

3. What do you need to do to improve your recall process?
Check one: organize___ rehearse___ use___ recall OK____

POSITION PAPER ON THE ROLE OF CALCULUS IN THE EDUCATION

OF STUDENTS IN THE BIOLOGICAL SCIENCES

for the

New Orleans Workshop, January 2-6, 1986

by

H. R. van der Vaart
Graduate Program in Biomathematics
Statistics Department and Mathematics Department
North Carolina State University
Raleigh, North Carolina

INTRODUCTION

My present contact with the problems under discussion is that in

teaching a first year graduate course in Biomathematics I find in my classes

a goodly number of Biology majors (in addition to some Mathematics majors,

some Statistics majors, of course Biomathematics majors, and even some

Chemical Engineers). Among these Biology majors there are some who have

taken extra Mathematics courses during their undergraduate years, as well

as some who have not.

In earlier years, I have taught an experimental course in introductory

mathematics to a somewhat select group of undergraduate Biology students.

This course included, in addition to much of the traditional calculus material,

topics such as difference equations and a little bit of linear algebra, and

emphasized the concept of mappings as unifying thread throughout the course.

Some students who took part in that experiment, reminisced years later that they

had discovered the beauty of mathematics in that course; others, I think, found

it a bit hard going. I also taught calculus in a Summer Institute at the

Uninversity of Michigan to a group of research biologists who, at a more mature age,

wanted to learn more mathematics. In that summer institute I used my own notes and

the book by S. Lang (1964). Several participants related that after the course they felt entirely comfortable with Lang's book and enjoyed referring to it later.

In these courses I did not emphasize applications as much as the simplicity and essential unity of the fundamental ideas involved in the structure of calculus: mapping, composite mapping, linear mapping, local approximation of nonlinear mappings by linear ones. It must be noted that computing had not made its big splash yet when I taught these courses.

Finally, I was a member of the Panel on Mathematics in the Life Sciences of the CUPM.

Now to come to the matter at hand: before discussing the calculus course, the instruction, I want to cast a long glance at the students, the instructed. I will do so in three parts:

a) Where do the students come from? their pre-college years.

b) Where are the students when they take the course? their college environment.

c) Where are they going? their future needs.

PRE-COLLEGE BACKGROUND OF THE STUDENTS

No longer does one have to fight an up-hill battle to convince people that many students of high school age are tremendously short-changed as to their intellectual training, including mathematics. Nor is this typical for just the future biology students. Two consequences of the situation are relevant to our present discussion, however, and the second one of these *is* peculiar to biologists.

α) Because of the existence of some exceptional high schools, including some private prep schools, which do perform outstandingly, and because of some schools doing a middling job, many colleges are faced with a wide range in preparedness of the biology freshmen that present themselves for their two bits of mathematical training; the width of this range is enhanced by the second circumstance:

β) All too many high school counselors tell a student who is weak in

mathematics, but professes an interest in "science", to go to biology, "since

you will not need much math there".

THE COLLEGE ENVIRONMENT OF THE STUDENTS

Several of the boundary conditions under which calculus, or any topic in

mathematics for that matter, is taught to biology students, are not under the

control of the mathematics department; at least, such is the case at many

schools. It might be useful to contemplate what (if any) changes in this

respect would be feasible in view of campus politics; it might also be useful

to find out what differences there are in this respect between schools. I am

referring to such circumstances as the following:

i) The above-mentioned variety in preparedness on the one hand necessitates

remedial courses for the weak, on the other hand opens the possibility for

stronger courses for the strong. On both scores we may run into controls

exerted by the biology departments:

ii) The major department usually decides which courses shall be counted

toward fulfillment of graduation requirements. At my University we are in a

relatively good situation: The Biological Sciences do not count "Algebra

and Trigonometry" toward graduation. However, the Undergraduate Catalog

shows that the Psychology department allows "Algebra and Trig" to be so

counted, and that the Sociology department lists it as a regular part of

the curriculum. I would not be surprised if there are schools where the

Biological Sciences have a similar policy. Let me add that my Undergraduate

Catalog also shows that Psychology does not require any calculus to be

taken (just 2 math courses) , and that Sociology lists the first semester

of a 2-semester calculus package as its second (and final) math requirement.
However, Sociology also requires an introductory Statistics and an introductory
Computer course, while Psychology even has two Statistics courses and one
Computer course. As for my Biological Sciences departments: the catalog lists
2 calculus courses designed for the softer sciences plus one course to be
chosen from Statistics, Computer Science, or Mathematics (subject to approval) ,
or the three-semester calculus sequence for engineers. It is here that on my
campus (and undoubtedly many other campuses) we run into another control
exercised from outside the Mathematics department.

iii) The student's biological advisors routinely argue strongly that
even the well-prepared student shall not take the more rigorous calculus
sequence. They present such action to the student as an unnecessary risk:
"You don't need it; you risk getting a lower grade, and see what that will do
to your record". Several of these students are pre-med: one reference to
low grades is enough to dissuade them. A teacher I know prevailed in the
face of such opposition, upon a mathematically strong biology student in her soft
calculus class to switch to a hard calculus class. Wouldn't you know it : he
hit a lousy teacher and hard grader (there are such!) . I am sure that his
example will be quoted by many biology advisors for years to come. So much
for the opportunity for well-prepared biology students to take stronger
calculus (or math) courses.

iv) The biology students rarely see an application of calculus in their

indergraduate biology courses. Some statistics, yes, but hardly any calculus.
It is easy and perhaps a bit cheap to blame this on the biology professors.
There are, in fact, at least two reasons for this fact which both stem from
the nature of biology and are virtually uncontrollable by either biologists
or mathematicians. One: the primary task of the biology curriculum must be

to initiate the students into characteristically biological concepts, the
biological way of thinking, and (obviously) a number of basic results
(mostly obtained by experimental methods) , as well as into a number of
basic laboratory techniques. In view of the large number of fields of inquiry
within biology, many with very specialized conceptual structures, this is a
huge task indeed. None of this work will gain much from mathematical intervention:
calculus played hardly any role in the development of these concepts and to
use calculus secondarily in their exposition or in the exploration of their
immediate consequences will often be genuinely impossible (e.g., think of
such fields as anatomy or taxonomy with their many category-type concepts)
or, at best, look like a somewhat artificial device to make some idea in
calculus more palatable, rather than elucidate new (to the students) and
tricky ideas in biology. Although mathematical methods have now entered some
of the more mature areas of research in biology (e.g., genetics, ecology,
(neuro-)physiology and biophysics), this research level of biology is much
more apt to be encountered in graduate studies then in undergraduate training.
Two: whenever mathematical methods have been usefully applied toward the
solution of biological problems, they have rarely been derived from plain
calculus: most such applications are either non-calculus (dimensional analysis,
graphical methods, inequalities, difference equations, pre-calculus probability,
linear programming, linear algebra) or post-calculus (ordinary differential
equations, partial differential equations, functional differential equations,
post-calculus probability) . The most prominent exception to this trend
seems to stem from those optimization problems that can be formulated as the
determination of the extremum of a function of one or more variables. Such
optimization problems, however, do not occupy center stage in the undergraduate biology curriculum. For examples of applications of the above-
listed areas in mathematics the reader is referred to such books as Gold (1977),
Marcus-Roberts and Thompson (1983), Noble (1967), Bender (1978), where

further references can be found.   Even an elementary textbook

such as Batschelet (1976), which has some realistic examples in several of

its chapters, has pretty lame examples in its calculus-related chapters.

In the present context it would be of obvious interest to learn from

biology faculty which biological problems they might be interested in seeing

treated mathematically during their undergraduate curriculum.   To whet your

appetite, let me give you one such list that came up during a project at

NCSU in which some people in our Biomathematics Program and some people of

the Biology faculty were exploring ways of injecting some mathematics into

the introductory biology course.   The biologists said they would be interested

in self-supporting 10-15 minute modules in each of the following topics:

α)   Enzymatic reactions ;   basic enzyme kinetics

β)   Cellular respiration ;   product quantity as related to
availability of reactants and rates of reaction

γ)   Photosynthesis ;   limiting factors ; C_3 and C_4 carbon
fixation

δ)   Transmission of nerve impulse ;   resting and action potential

ϵ)   Countercurrent flow mechanism ;   loop of Henle in human kidney

ζ)   Blood circulation :   blood pressure, stroke volume of the
heart, diameter of blood vessels, distensibility of arteries

η)   Ecology :   growth curves, carrying capacity, life tables.

I am not sure that one can do justice to any of these topics in the short

time indicated or, for that matter, in an introductory course.   But as an

indication of what biologists might be interested in gaining from mathematical

methods, I think this list has great interest; and I think we should solicit

more such lists from more biologists in various places.   Before leaving the

topic of calculus applications in biology courses I want to make two remarks

regarding the topics in the list:

1) The preparation of these topics for mathematical discussion will require more than just labeling certain quantities by x , y , and z and plunking down an equation to be solved : a real modeling effort must be made where biological or biochemical concepts are exhibited first, and where it is then discussed how to represent mathematically which of their features.

2) The derivation of the properties of the resulting mathematical models for these biological situations will require a lot more than simple calculus.

v) One might hope that the biology students might see some applications of calculus in the <u>physics</u> or <u>chemistry</u> courses they are <u>required</u> to take. But alas! On many campuses the biologists take non-calculus (or very-little-calculus) versions of these courses. At NCSU the Math prerequisite for their year in Physics is Algebra and Trigonometry <u>or</u> Finite math, and the textbooks in the listed Chemistry courses go to great lengths indeed to avoid any use of calculus whatsoever. In fact, on my campus the first two chemistry courses that chemistry majors take, avoid the use of calculus.

vi) Another potential contact between biology students and calculus, viz., the introductory <u>statistics</u> course, does not usually work out either: these courses are traditionally light on the calculus, in fact, light on the use of any mathematical formulae, and they are now developing in the direction of including the teaching of the use of statistical computer packages. The same is true for possible courses in the <u>simulation</u> of biological systems: canned programs, written in some simulation language are the order of the day in such courses.

THE POSTGRADUATE LIFE OF THE STUDENTS

The last question we want to consider with respect to the students is
this one: which fields of mathematics could they run into when they are in
graduate school and when they wind up doing research?

One thing we may be sure of is that they will be applying some elementary
statistical methods, mostly because those are needed to analyze their
experimental data, but sometimes simply in order to have their papers accepted
by this or that biological journal: many editors insist on the use of
statistical methods, even when (unbeknownst to them) such are not really
needed.

If their research calls for complicated experiments where many variables
are being measured and where the goal is to find out how these variables depend
on several influencing factors, then they will probably need multivariate
statistical methods, possibly nonlinear methods; so they will have to reach
for much more sophisticated methods. Thus they need some additional training
in statistics, in order that they can talk more fruitfully to statistical
consultants, or be more efficient in choosing which part of what statistical
computer package will answer their needs: part of their training has to consist
in learning how to do this. One aspect of all of this that should worry
us here and now, is the fact that this statistical software has not only
statistical content, but also some (numerical) mathematical content. For
instance, in any least squares method there are matrices to be inverted. These
matrices are not always well-conditioned. For a long time the most popular
method among statisticians for matrix inversion has been the Doolittle method,
since it had been worked out so that it gave a lot of extra statistical
information while the inversion process was going on. However, the Doolittle

method is one of the methods that are numerically least stable. The question

thus arises: is a particular piece of statistical software safe to use for

a particular concrete problem? Does this problem harbour a problem matrix?

Can this matrix be inverted safely by any method? by the method used in this

particular computer package? This little example illustrates a major headache

of the so-called computer age:

> Is it possible (and if so, how?) to train students to be critical
>
> of canned programs, so that they know what kind of pitfalls to expect
>
> and may be better able to find out if a given program might serve them
>
> well, or whether they should look for help if the results are unsatisfactory,
>
> or how to choose another program?

Nor is this problem restricted to statistical packages. Kahan (1972) ,

pp. 1227-1229, relates what he himself calls a 'horror story', according to

which a candidate for a Ph.D. degree in aeronautical engineering did <u>not</u>

see his new model for <u>improved</u> wing lift confirmed by the computer output

from his new equations:

> in single precision (as it turned out) because at that time IBM's
>> single precision logarithm subroutine was flawed,
>
> in double precision (as it turned out) because the double precision
>> subtraction hardware on their particular IBM model lacked a guard
>>> bit; he got around this deficiency by a clever programming trick.

This story is an example where the blacksmith's iron deserves all the blame

the blacksmith is willing to heap on it. The obvious problem is: how does the

user know when to believe the computer and when not? How does he know when

to blame the program and when the machine? These problems are aggravated

when the errors are in the hardware or in proprietary software. Nonetheless,

the above Ph.D. candidate was lucky in that he chose to go to W. Kahan for

help and in that W. Kahan just happened to work on some related problems
(l.c.) so that it did not take him too long before he started uncovering
the weak spots in the situation.

Beyond their need for statistical savvy and a certain type of computing
knowledge there is little that can be said about general mathematical needs
of biologists. Especially when a person specializes in Biomathematics,
in Theoretical/Mathematical Ecology, in the physiology of organ systems such
as heart, lungs, kidney, brain, nervous system, or in Insect Pest Management,
or in models for crop growth, individual plant growth, or photosynthesis,
there are very few types of mathematics that one would want to guarantee that
person will never get involved with. Certainly numerical simulation will
hit him sooner or later, but among the more traditional fields in mathematics
chances are excellent that there will be contact with systems of (nonlinear)
difference and differential equations (ordinary and partial), various
concepts of stability, reaction-diffusion models (as an entry into spatio-
temporal phenomena), stochastic versions of any of the above, systems with
more than one time scale (for bio-chemical dynamics, for theories that
connect ecology and evolution, or ecology and physiology), techniques
borrowed from engineering (chemical, electrical, mechanical), artificial
intelligence, to name just a few.

SOME THOUGHTS ABOUT UNDERGRADUATE MATHEMATICS FOR BIOLOGISTS

Obviously, a few undergraduate mathematics courses cannot prepare
anybody (including biologists) to cope with all these areas. Indeed, nobody
who is going to use mathematics for much of his professional life has the right
to expect being equipped forever by a little bit of undergraduate work.
Mathematics majors take a lot more mathematics before their bachelor's degree,
and they are far from finished when they receive that degree. What might

perhaps be feasible is to give the biology students a framework that they
will not have to unlearn later, but instead can build on. For instance,
geometric and qualitative phase plane methods are much more important for
their introduction to ODE's than are mechanical tricks for the integration
of special types; the intuitive feel for how the solutions go from initial
value at t_o to current value at t is more important than games played
with integration constants. Similarly they should not be taught things they
will never use (e.g., centers of gravity, moments of simply shaped objects) ,
nor things that will turn them off from mathematics (they will despise
"applied" problems such as "Suppose a population grows according to the
formula $x_n = 1000 + 500(1 - 2^{-n})$") Memorization of a large number of
derivatives and indefinite integrals will not do them nearly as much good
as the actual derivation (not necessarily a rigorous proof, but an argument
that illustrates the concepts) of a few well-chosen examples. Use of
graphical methods is strongly recommended.

It is hard to propose a list of topics to include. A list might easily
become too long. But certain themes reverberate through much of the mathematics
that one encounters in the biological research literature. The following
are worthy candidates.

Computer problems : error analysis, floating point numbers, calculator
math, etc. ; do not put a 'blind trust' in your computer! ;
conflict between increasing round-off and decreasing discretization
errors.

Continuous vs. discrete models : some parallel development of
difference and differential, sum and integral calculus.

Different behavior of solutions to certain differential equations

and "similar" difference equations. (I have had two personal
encounters, separated by many years, with mathematical biologists
that were not aware that asymptotic stability is not necessarily
preserved when one discretizes an ODE) .

Basic features of linear mappings, linear algebra, composite of linear
mappings, local approximation of nonlinear objects by linear ones.

Basic concepts like mapping, inverse (not necessarily unique) ,
composite ; many mathematical models are of form $L x = a$,
"Fredholm alternative" for system of linear equations.

Rate of change (e.g. Feynman's vol. 1 of lect. on physics, pp. 8-3, 8-4 :)
derivative, linear local approximation, mathematical examples to
practice the concept ; higher order local approximation (Taylor) .

Obviously I doubt the wisdom of separating calculus and calculator, or even
calculus and computer. Who is to do the programming instruction I do not
know. I do think the mathematics department should teach a sound dose of
error analysis, e.g., some of the stuff that has recently appeared in the
Monthly on "calculator math" .

REFERENCES

Batschelet, E. (1976), Introduction to mathematics for life scientists;
New York, Springer, 2nd ed., xv + 643 pp.

Bender, E. A. (1978), An introduction to mathematical modeling; Wiley-
Interscience, x + 256 pp.

Feynman, R. P., Leighton, R. B. and Sands, M. (1963), The Feynman lectures
on physics, vol. 1, 12 pp. + 52 chapters + 3 pp.

Gold, H. J. (1977), Mathematical modeling of biological systems; an
introductory guidebook; Wiley-Interscience, xv + 357 pp.

Kahan, W. (1972), A survey of error analysis; pp. 1214-1239 in:

    C. V. Freiman (ed.), Information processing 71, Proc. IFIP Congress

    71, vol. 2, applications; Amsterdam, North-Holland, xv + 1618 pp.

Lang, S. (1964), A first course in calculus, xii + 258 pp; A second course

    in calculus, xii + 242 pp.; Mass., Reading, Addison-Wesley.

Marcus-Roberts, H. and Thompson, M. (eds.) (1983), Life science models;

    xx + 366 pp; vol. 4 of: W. F. Lucas (ed.), Modules in applied

    mathematics; New York, Springer.

Noble, B. (1967), Applications of undergraduate mathematics in engineering;

    MAA; N. Y., Macmillan, xvii + 364 pp.

A SENSIBLE APPROACH TO CALCULUS

by

Carol Ash
Robert Ash
M. E. VanValkenburg

Engineers often lament the fact that mathematics courses are so ineffective in teaching engineers how to use mathematics. Similar statements appear in a book to be published soon.

"Mathematicians and consumers of mathematics (such as engineers) seem to disagree as to what mathematics actually is. To a mathematician, it is important to distinguish between rigor and intuition. To an engineer, intuitive thinking, geometric reasoning and physical deductions are all valid if they illuminate a problem, and a formal proof is often unnecessary or counterproductive.

"Most calculus texts claim to be intuitive, informal, and even friendly, and in fact one can find many worked-out examples, as well as some geometric and physical reasoning. However, the dominant feature of these books is formalism. Definitions and theorems are stated precisely, and many results are proved at a level of rigor that is acceptable to a working mathematician. We admit to a twinge of embarrassment in arguing that this is bad. However, our calculus students have ranged from close to the best to be found anywhere, to far from the worst, and it seems entirely clear to us that most students are not ready for an abstract presentation, and they simply will not learn the formalism. The better students will succeed in reading around the abstractions, so that the textbook at least become useful as a source of examples."

227

The authors of this statement are Robert and Carol Ash. Robert was educated as an electrical engineer and he taught EE before transferring to the Mathematics Department. The book titled "The Calculus Tutoring Book," will be published late in 1985 by IEEE Press.

Traditional publishers of books on calculus maintain a priesthood of advisers who would never let such a book pass. But a friendly reception by electrical engineers is assured. Where will it be used? For universities with an engineering-oriented mathematics department, the book might well be used as the class textbook. Failing this, it can be used for supplementary reading for both calculus classes and for courses in engineering.

Based on their twenty years of teaching experience, the authors write:

"Our approach uses informal language and emphasizes geometric and physical reasoning. The style is similar to that used in applied courses, and for this reason students find the presentation very congenial. They do not regard calculus as a strange subject outside their normal experience."

Here, in the words of the authors, is a more detailed discussion of the philosophy and style of the Calculus Tutoring Book -

We teach in the Mathematics Department at a Big Ten School with a large and selective engineering college. In many math courses, such as calculus, differential equations and linear algebra, the majority of students are in engineering but the instructors, consciously or unconsciously, present the subject as if the entire audience were planning a career in pure mathematics. When we started teaching engineering mathematics, we too stated

careful definitions and proved theorems. But we found that our formal
mathematical language which was intended to prevent misunderstanding had
precisely the opposite effect. It obscured ideas that students would
otherwise find straight forward. For example the formalism

$$f(x,y,y') = 0$$

disguises the simple idea of an equation involving x, a function y and its
derivative y', such as $3x + 2x^2y^4 = 5y'$ or $y' = x^2y$. We also found that
typically after proving a theorem, we would be asked, "Why is it really
true?." Even the few who followed every line of a proof simply were not
<u>convinced</u> by it. Disaster was avoided only because students learned to
politely ignore our proofs and definitions in class and we learned to
diplomatically avoid them in making up exams. The net result was that our
courses were much less than they could have been since the intellectual level
is not determined by what we self-righteously do at the blackboard but by what
we actually get students to be able to do as a result.

Changing our style

A mathematician views a proof as a logical progression from a set of
hypotheses to a conclusion using definitions, axioms and rules of inference.
Physical and geometric reasoning, possibly with the help of diagrams or
intuition, is a useful aid but is not acceptable as a legal part of the
argument. On the other hand, a physicist or engineer regards the underlying
physics, geometry and intuition as the core of the problem. Mathematics is
useful as an aid to the understanding, but it may be stretched and otherwise
manipulated, in ways that a professional mathematician would find
unacceptable, if this leads to a sharper insight into the physical situation.

We believe that the physical and geometric way of thinking is more valuable when seeing a particular subject for the first time, even if the student eventually becomes a mathematics major. Mathematics does not come into existence fully developed with theorems and proofs. It arises from imagination and intuition aided by physical and geometric reasoning. Students should be taught in a style that reflects the creation of mathematics and not in style that would satisfy a professional mathematician tidying things up years after the fact. It is more important to learn how to formulate and solve interesting problems than to learn the techniques of writing formal proofs.

Over the years our teaching approach has evolved until we now use relatively informal language, and stress underlying physical and geometric ideas. We give explanations for the ideas and procedures of engineering mathematics, but they are not necessarily in a form suitable for publication in professional mathematical literature. Instead, they reflect how we ourselves actually think. For example, we remember the theorems of calculus not because we have seen them proved formally but because each says something about slope or area or velocity, etc. that seems reasonable. We try to explain why a result is "really" true and give students a way of thinking so that they can learn to decide on their own what is true. Students now leave our courses with their initial reserve of good will and self-confidence intact. We feel that those who take later courses in pure mathematics are better served by this earlier experience than they would have been by a premature exposure to formalism. Those who continue in a standard engineering

program are also better served since in their own fields, they must <u>apply</u>
mathematics (using geometric and physical reasoning) and not construct proofs
to meet the standards of professional mathematicians.   One of our students
approved of our approach because "she taught in English, not in Mathematics".

The Calculus Tutoring Book

We have written a calculus text (<u>The Calculus Tutoring Book</u>, IEEE Press,
1985) using our approach. We'd like to give a few instances of how it differs
from traditional texts.

The chain rule for derivatives states that -

$$(1) \qquad \frac{dy}{dx} = \frac{dy}{du} \, \frac{du}{dx}$$

We see why it is true, as follows: If say dy/du = 3 and du/dx = 2, then
y is changing three times as fast as u, while, in turn, u is changing twice as
fast as x. So all in all, y is changing six times as fast as x, i.e., dy/dx
is the product of dy/du and du/dx, as stated in (1).

In a typical text, first the theorem is stated precisely:

Let y = f(u) and u = g(x) where the derivative of g exists at x

and the derivative of f exists at g(x). Then the composition $f \circ g$ is

differentiable at x, and (1) holds.

Then a proof is given, along the following lines:

Suppose x changes by Δx. Then u changes by Δu. Furthermore

(2) $\Delta u = \dfrac{\Delta u}{\Delta x} \ \Delta x$

so as $\Delta x \to 0$, the factor $\Delta u/\Delta x$ in (2) approaches the

number du/dx and the entire product, namely Δu, approaches 0.

Now consider

(3) $\dfrac{\Delta y}{\Delta x} = \dfrac{\Delta y}{\Delta u} \ \dfrac{\Delta u}{\Delta x}$

Allow Δx (hence Δu) to approach 0 in (3) to obtain the chain

rule in (1). But a technical difficulty arises because

although it is assumed that $\Delta x \neq 0$, it is entirely

possible for Δu to be 0, in which case, (3) contains an

illegal division by 0.

At this point, some authors push on to give a complete proof, and others

stop after at least noting the difficulty. In any event, very little has been

revealed about why the chain rule is valid. Although abstract mathematicians

tend to dismiss our version of the proof as "imprecise" or "sloppy," it is in

fact consistent with the way mathematics is actually used in basic physics and

engineering courses, and therefore much more appropriate for the beginning

student than the formal version.

As another example, consider L'Hospital's rule which says that if

$$(4) \qquad \lim_{x \to a} \frac{f(x)}{g(x)}$$

is of the form 0/0 then we try to find the limit by switching to –

$$(5) \qquad \lim_{x \to a} \frac{f'(x)}{g'(x)}$$

Most texts prove the rule using the Extended Mean Value Theorem which in turn is proved using the Mean Value Theorem which in turn is proved using Rolle's Theorem which in turn is proved using an Extreme Value Theorem whose proof ironically is widely admitted to be beyond the scope of any calculus text. Again, this approach yields little insight. It does not even explain why the form 0/0 is referred to as indeterminate and requires a special procedure.

Our approach is to begin with some numerical instances of the limit form 0/0 such as those in Tables 1-3. In each case, the numerator and denominator approach 0 but they do so at different rates and, therefore, produce different limits. In Table 1, the fraction has limit 0 (the numerator approaches 0 much more rapidly than the denominator); in Table 2, the fraction has limit 2; in Table 3 (where the denominator approaches 0 much faster than the numerator) the fraction has limit ∞ . By comparison, any problem say of the form 3/0+ has the answer ∞ no matter how fast the numerator approaches 3 and the

Table 1

numerator	.1	.01	.001	.0001	...
denominator	1	1/2	1/3	1/4	...

Table 2

numerator	2/3	2/4	2/5	2/6	...
denominator	1/3	1/4	1/5	1/6	...

Table 3

numerator	1/2	1/3	1/4	1/5	...
denominator	.1	.01	.001	.0001	...

denominator approaches 0 from the right. When two problems of the same form can have two different answers, we call the form indeterminate. The form 0/0 is indeterminate (the answer depends on the particular numerator and denominator) while the form 3/0+ is not indeterminate (all such problems have the answer ∞).

Now that 0/0 has been singled out as special, we can outline the intuitive idea behind L'Hospital's rule. Corresponding to a numerator and a denominator both approaching 0, think of two runners at a starting line (the zero mark). At that moment, the ratio of their positions is the arithmetic impossibility 0/0. But as they move, the ratio of their positions near the starting line, where they began "even with each other" depends on how fast they move. For example, if the first runner is going twice as fast as the second and they cross a zero line together, then very near that line, the first runner is twice as far from the line as the second. So it makes sense to replace the ratio of <u>positions</u> f(x)/g(x) in (4) by the ratio of <u>velocities</u>

f'(x)/g'(x).

Sensible calculus meets Mr. Ugh

When the first draft of <u>The Calculus Tutoring Book</u> was subjected to the

standard publishing process, we were reviewed by a group of abstract

mathematicians. They objected strenuously, even to minor simplifications in

notation such as $e^{-\infty}=0$ as an abbreviation for $\lim_{x\to-\infty}e^{x}=0$.   Our informal

arguments were grudgingly tolerated but only under the condition that we

provide appendices with complete legal proofs. One reviewer expressed his

opinions by decorating our manuscript with "ugh" whenever he was displeased.

Another actually commented that "the students must be given the facts [meaning

theorums and proofs] even if they can't understand them". We agreed to write

appendices but as the process continued, the appendices got longer and longer

until our original idea of writing an informal version of calculus was

engulfed by clouds of abstraction. At this point, we realized that there was

no way to satisfy the reviewers, and that the publication of the book would

have to be unorthodox. We are grateful for the friendly reception we have

received from the IEEE. We hope that students will find the text useful and

that it will be a step in the direction of more appropriate teaching of

mathematics.

APPENDIX

From the Ivory Tower

IEEE-ASEE NEWSLETTER

Spring issue, 1985

Gerald R. Peterson
Editor

For the first time in many issues we are missing Mac Van Valkenburg's Curriculum Trends column. Mac underwent major heart surgery in January. Last we heard, he's doing fine, but not yet up to producing a column for this issue. I know all his many friends wish him a speedy recovery, and we look forward to the return of his informative and thought-provoking column in the next issue.

In several recent columns, Mac has commented on the deplorable state of instruction in college-level mathematics, with special reference to the fact that most math teachers at the college level seem to be either unwilling or unable to teach math in a way that makes it relevant or meaningful to anyone except professional mathematicians. His comments definitely struck a chord with me. The math department here at Arizona is probably more coop-

erative than many -- they periodically ask representatives from engineering to serve on calculus text-selection committees. But it is only a ceremonial courtesy. We faithfully review several dozen candidates, picking out those that seem to make some effort to relate mathematics to the real world, but our choices are invariably rejected as being "not sufficiently rigorous." Well, this will come as no news to most of you, but I am sorry to report that this deplorable state of affairs seems to be extending into the high schools.

My daughter is a senior in high school this year and took College Algebra last fall. Some of you may object that this isn't quite fair since it is "college" algebra, but the fact is that it is routinely taught in high school and is, indeed, an entrance requirement for most engineering programs. The text was a stan-

2

dard college algebra text, by a well-known author of college-level math books, whose name many of you would recognize. In this book, you will find the following definition:

If S is a relation:

$$S^{-1} = \{(x,y) \mid (y,x) \in S\}.$$

Many of you took your PhD. minor in math, so you're experts -- what does that mean? It's the definition of an inverse function. It can be stated in plain English. If y is a function of x, the inverse is obtained by replacing y with x and x with y wherever they occur. But does the book ever say that? Need you ask? Well, whatever you may think of the value of exposing the student to mathematical formality, the most serious fault, in my view, is that the author never gives the slightest hint as to why this inverse function might be of interest to anyone. He presents it, with just the formality I have indicated above, gives a few problems for the teacher to assign, and goes right on, never mentioning it again.

Well, maybe that's an isolated horrible example. Or is it? My daughter's boyfriend is taking College Algebra this spring. He has a different text. (There's another nasty habit the high schools seem to be picking up from us, changing texts every time the instructor changes.) Since the other text was so poor, we might hope

for an improvement. Suppose I tell you that, given the polynomials:

$$4x^2, \quad 15x^3y, \quad \text{and} \quad 36xy^2,$$

the least common multiple is $180x^3 y^2$. Does that mean anything to you? Given a set of polynomials, the least common multiple is simply the smallest polynomial into which all divide evenly. Fine, but does the author ever say that? Need you ask? He never defines it, just gives a very complicated procedure for computing it, one that took me about 15 minutes to figure out, given the above polynomials and the answer. Maybe I'm just dumb, but again the biggest problem is that the author never gives the slightest hint as to why anybody in his right mind (that excludes mathematicians) should have the slightest interest in a least common multiple.

I find this very discouraging. I have had the notion (probably naive) that high school math teachers might actually be interested in teaching math. After all, they don't have to publish any incomprehensible research papers to keep their jobs. Maybe they figure they should use books like this to prepare students for the horrors of what they will face in college math classes. But I wonder how many students they turn off to mathematics before they even get to college. What can be done about it? If you have kids in high school, you might check into what is going on in their math classes and perhaps offer a suggestion or two if you think useful changes could be made.

Computer Symbolic Manipulation in Elementary Calculus

Paul Zorn, St. Olaf College

Northfield, Minnesota 55057

Introduction

Although numerical computing has been available for over twenty years, it has so far hardly affected the shape or content of the mainstream college calculus course. The advent of powerful computer algebra systems (such as SMP, MACSYMA, and MAPLE) will be harder to ignore. Symbolic manipulation programs (hereafter, "smp's") have two important new features: first, they handle nearly every routine calculus operation--symbolic algebra, formal differentiation and integration, expansion in power series, numerical computations (as of Riemann sums), and graph sketching; and second, they do all of this without programming. By assuming some of the burden of manipulation, smp's can help students see the richer ideas beyond. In particular, by making possible a better balance between discrete and continuous viewpoints, explicit and implicit functions, and exact and approximate methods, smp's can help broaden calculus's scope and show more of its power.

Calculus as it is taught

Which problems can smp's help to solve? What is wrong with courses as they are? Are the ideas of calculus outmoded, unimportant, useless, or uncompetitive with discrete mathematics?

Before complaining about how calculus is taught, one must affirm that even limited to the elementary functions, "exact" calculus is a beautiful, coherent set of ideas that is adequate to solve many classical problems. The

main _ideas_ of calculus are as important now as they have ever been. If not
more so--the rise of discrete mathematics should boost calculus, not kill it.
Analysis of algorithms, for example, offers genuine applications of calculus
that deserve to appear in calculus courses alongside classical problems from
physics. Such applications demonstrate the power of calculus beyond its
traditional borders, and they illustrate what should be one of the main themes
of both calculus and discrete mathematics: the interplay of discrete and
continuous ideas. Discrete mathematics will generate new demand for calculus,
but not for the standard compendium of closed-form techniques applied to
elementary functions. Discrete mathematics students need to understand such
basic calculus _ideas_ as limits, rates of change, and qualitative growth
behavior.

Calculus' most serious problem is that its good, old, important _ideas_
have _given_ _way_ _to_ _techniques_, some of which are indeed outmoded, unimportant,
and uncompetitive with discrete mathematics. In olden days students began
calculus with good technical preparation (especially in algebra). They could
see beyond the formal operations to the questions those techniques address and
the ideas they illustrate. Now most students calculate poorly, and calculus
has become by default a course in calculation. Formal differentiation,
methods of antidifferentiation, limit computations, and convergence testing--
usually applied to explicit algebraic objects--take most of the class time and
appear most dependably on assignments and tests. Some techniques are entirely
routine and some are not, but all amount to performing circumscribed tasks in
explicitly prescribed ways. In an essay [3] on mathematical maturity, Lynn
Steen accuses calculus of "programming people to serve as moderately
sophisticated computers." Instead of learning to create, verify, and analyze
algorithms, calculus students learn to perform them.

Undue emphasis on techniques leads to another problem: calculus courses focus too narrowly on closed-form methods applied to elementary functions. The standard formal techniques apply most conveniently to explicit algebraic combinations of elementary functions, but relying too much on these examples obscures the ideas they illustrate. In effect, it turns calculus into algebra. Students often mistake the chain rule, for example, for an algebraic rather than a functional equation. We foster this misunderstanding by relying too much on elementary functions: applied only to them, the chain rule looks like a prescription to do some algebra. We could force students to encounter the idea of the chain rule (and many other ideas in calculus, for that matter) by routinely considering other-than-elementary functions: functions given graphically, implicitly, recursively, by tables, or in "black box" form. Such functions sometimes appear in calculus as examples (usually pathological!), but rarely in exercises or as objects of study in their own right. Ironically, most of the functions students will see in more advanced work are to some extent inexplicit; explicit algebraic functions then arise mainly as idealized examples.

Calculus chooses its techniques, like its functions, too narrowly. Compared with "exact" or "closed-form" methods, approximate, numerical, iterative, and recursive techniques get much too little attention. Numerical integration, for instance, is in the book, but usually segregated in an "optional" section. Numerical solution of equations may not be treated at all. Taylor series, which should motivate studying power series at all, too often follow a long, gruelling ordeal with convergence tests. By then, students don't consider Taylor series a reward for anything.

Because closed-form methods fail with so many innocuous-looking problems (minimize $x^2 - \sin(x)$ on $[0,1]$), exercises and applications are carefully

contrived, and they show it. Arc-length integral problems are especially
ludicrous: they can almost never be computed in closed form. An excellent
opportunity to use numerical integration where it is needed is usually wasted.
In the calculus text I use, numerical integration is still four chapters away.
Having learned only exact techniques, students naturally conclude that
calculus solves only artificial problems, and they doubt their own ability to
use it. If exact and approximate methods were treated together in calculus,
students could solve more problems, and more interesting ones, with less
data-censoring. They would better appreciate calculus's versatility. Most
important, they would see in action that crucial relationship between
"discrete" and "continuous" which so usefully undermines the belief that
calculus is algebra.

As with its functions and its techniques, the range of calculus
applications is too narrow. Simple physical applications are convenient
because they often lead to elementary functions and exact techniques.
Modeling other phenomena may require discrete methods and lead to more
interesting mathematics. Problems of finance, for example, look difficult for
traditional calculus because their variables tend to be implicit functions of
each other. Equations relating the variables, which are often polynomials of
high degree, need to be solved numerically. When this can be done, such
problems become elementary applications of (polynomial!) calculus.
Computational complexity is one of many important examples that come from the
theory of algorithms and computer science. Unfortunately, few texts even
mention the big-oh formalism.

Discrete and approximate viewpoints have been slighted, but for an
excusable reason: studying other-than-closed-form methods honestly in
calculus requires considerable numerical and algebraic manipulation. The work

required is mostly tedious and meticulous but straightforward, and often beside the main point. (In other words, precisely the sort of work that ought to be left to a machine. Fortunately, that is now possible.) Consider, for example, Simpson's rule for approximating a definite integral. To compute the approximation at all is tedious and distracting unless the integrand is simple and the number of subdivisions small. Any computer or programmable calculator will do the numerical work, but the problem of estimating the error remains. The standard error formula involves maximizing the absolute value of the fourth derivative of the integrand over the interval of integration. For any but the simplest integrands, doing this "by hand" is time-consuming or impractical. For simple integrands, the task is artificial: such integrals may as well be evaluated in closed form. As the example session in the appendix shows, a powerful smp makes the problem routine.

What to do, and how smp's can help

Some of the impetus for rethinking calculus courses stems from new interest in discrete mathematics. For the last two years, St. Olaf's mathematics department, under a grant from the Sloan Foundation, has devoted much effort to introducing a popular and apparently successful discrete mathematics course at the freshman-sophomore level. One of our most difficult problems was making room in the curriculum for the new course. After many discussions, we rejected the alternative of shortening calculus as likely to lead even further toward a sterile collection of techniques. We disagreed on the relative merits of discrete and continuous mathematics, but we agreed that the connection between the two is worth making, for the sake of both. For calculus, this entails (in the long run)

    -- more emphasis on theory, but supported by much more varied exercises and examples

-- a <u>unified</u> treatment of exact and approximate methods

-- some coverage of error estimates (not necessarily rigorous)

-- attention to qualitative behavior of functions and classes of
   functions (e.g., polynomial vs. exponential growth, big-oh)

-- explicit study of algorithms; calculus applied to generate, compare,
   and analyze them.

Incorporating such ideas into calculus courses does more than pack another

worthwhile topic into an already crowded syllabus. It offers a truer picture

and a deeper understanding of the calculus itself.

   The most important goal of these reforms is to rescue the ideas of

calculus from its techniques. The most important idea to be rescued is the

relation between discrete and continuous. Doing only the traditional

computational exercises with an smp is worse than useless, because it simply

transfers the technique from human to machine. But used creatively, smp's

can allow new kinds of examples and exercises that let students go beyond

routine manipulations of elementary functions to the ideas they are supposed

to illustrate. With smp's, for instance, students can work through concrete

cases of general theorems. For Taylor's theorem, this could mean computing,

graphing, and tabulating values of several Taylor polynomials, and observing

that the errors are as claimed. For the chain rule, students could

investigate the graphical meaning for many pairs of functions, or look at

difference quotients for composites of functions specified only in "black box"

form. (It is easy to hide a function's definition in an smp's bowels.) The

fundamental theorem, to choose a uniquely important example, says that two

apparently different things are the same. The point is lost when every

example integrand is antidifferentiable in closed form. But if values of the area-under-the-graph function can be computed numerically, tabulated, and graphed, for many integrands, the real idea of the theorem is hard to miss. With flexible graphics, students can see that differentiable functions look linear at small scale, quickly (if dirtily) estimate extrema, roots, and integrals, and get a general feeling for a problem before deciding whether and how to attack it in earnest. Smp's encourage experiments, conjectures, and other active work with mathematical ideas. They don't prove theorems, but they can help students learn that theorems answer questions that students are capable of asking.

The advantage of smp's in calculus teaching is that they remove the computational obstacles to giving the discrete viewpoint its due. With this done, students can be weaned from elementary functions, exact methods, and exclusively physical applications, and started on a more varied and substantial diet. With its intellectual content restored, calculus can serve its dual purpose as a flexible tool and an introduction to mathematical thinking.

Calculus and numerical computing: some history

Smp's have exciting future prospects, but the lesson from history is discouraging. In the computer revolution, calculus has been a Tory.

Computers appeared in calculus in the early sixties, as soon as hardware and software improvements (BASIC, especially) made the idea remotely practical. Rationales for computing in calculus were issued from time to time by the Panel on the Impact of Computing on Mathematics Courses, an MAA subcommittee. In 1972 the Panel cited these benefits: more emphasis on algorithms; better motivation of approximation ideas; more realistic mathematical modeling; and better understanding of the process of problem-solving. Numerical computing was also supposed to generate student involvement and provide numerical illustrations of calculus ideas (such as convergence).

The most ambitious efforts to combine calculus with computing produced a few revisionist textbooks. The book of P. Lax, S. Burstein, and A. Lax [2], published in 1976, is an elegant example. Computing, numerical ideas, physical applications, and calculus theory inform and illustrate each other in this witty and iconoclastic treatment. Unfortunately, none of these comparatively difficult texts seems to be widely used.

Less radical efforts to introduce numerical computing into otherwise traditional calculus courses were commoner. In the late sixties and early seventies a spate of "computer supplements" to standard calculus courses appeared. These were generally collections of BASIC (sometimes FORTRAN) programming exercises. Typical programs compute Riemann sums or estimate derivatives numerically. One of the most interesting is Richard Hamming's Calculus and the Computer Revolution [1], published in 1968. Along with programming exercises, the book contains an eloquent discussion of the relation between computing and mathematics. Because Hamming emphasizes non-numerical algorithms and views the computer as a symbol-processing machine, his book still has a modern flavor.

Whatever its benefits may be, numerical computing has scarcely affected mainstream calculus teaching. In 1975, for example, a joint committee of the AMS and MAA issued a long report on various efforts to include computing in calculus, but also wondering (inconclusively) why there were so few. Computer lab courses associated to calculus do exist, and new textbooks contain calculator exercises and BASIC programs, but calculus itself has not assimilated the computing point of view. Calculus courses are still overwhelmingly closed-form techniques applied to explicit algebraic functions.

Why did numerical computer calculus fail to penetrate mainstream calculus? Why should smp's fare any better? The main obstacle for numerical computation was probably programming: even with BASIC, computer calculus was too much computing and too little calculus. The time and effort needed to program even simple calculus tasks was out of proportion to their value for learning calculus. At worst, students might find themselves without the mathematical sophistication to write the programs that would help them acquire the sophistication they didn't have.

Though they do much more, smp's handle traditional numerical calculations adequately for the purposes of a calculus student. What is important is that they do so interactively and without the distraction of programming. For example, the commands

```
Trapezoid[Cos[x 2],x,0,1,10]
Simpson[Cos[x 2],x,0,1,10]
Midpoint[Cos[x 2],x,0,1,10]
```

are all that a student using the smp SMP at St. Olaf College needs to compute the trapezoid rule, Simpson's rule, and midpoint rule numerical approximations to the integral in question. (See also the Appendix.) Output from any of the commands is always available to use elsewhere. Syntax must be learned, but not programming. All of the benefits claimed for numerical computing (not to mention the special possibilities of symbolic computing) are as valid as ever--with smp's, they can be realized.

Conclusion

Freshman calculus, like any other profitable business and giant bureaucracy, will not change overnight, whatever the logic of change may be. But if smp's find their way into calculus, they will certainly change it. Using smp's only for computational problems is foolish: it could cheat students out of that modicum of algebraic intuition that only practice can

develop. Using smp's should not deny the value of formal manipulation, but

help put it in its place, and buy time for better things.

    Calculus is largely the interplay of discrete and continuous ideas.

Smp's let us combine discrete and continuous ideas in a unified treatment that

can deepen student's understanding of both.

<u>References</u>

1. R. Hamming, Calculus and the Computer Revolution, Houghton Mifflin,
    Boston, 1968.

2. P. Lax, S. Burstein, A. Lax, Calculus with Applications and Computing,
    Springer-Verlag, New York, 1976.

3. L. Steen, Developing mathematical maturity, in The Future of College
    Mathematics: Proceedings of a Conference/Workshop on the First Two Years
    of College Mathematics, ed. by A. Ralston and G. Young, Springer-Verlag,
    New York, 1983

Appendix
Example Calculus Problems with SMP

A TRANSCENDENTAL EXERCISE

(Discuss $\displaystyle\int_0^1 \sin(x)\, \exp(-x^2)\, dx$)

#I[1]:: g[$x]::Sin[$x] Exp[-$x∧2]   (define g)

#O[1]:   ' Sin[$x] Exp[- $x^2]

#I[2]:: g[x]

#O[2]:   Exp[- x^2] Sin[x]

#I[3]:: Int[g[x],{x,0,1}]

#O[3]:   Int[Exp[- x^2] Sin[x],{x,0,1}]   (closed form integration fails)

#I[4]:: <'numint   (load a homemade numerical integrator)

#I[5]:: Simpson[g[x],x,0,1,4]   (Simpson's rule, 4 subdivisions)

#O[5]: $\dfrac{\text{Exp[-1] Sin[1] + 4Exp[-9/16] Sin[3/4] + 2Exp[-1/4] Sin[1/2] + 4Exp[-1/16] Sin[1/4]}}{12}$

#I[6]:: N[%]   (decimal form of previous line)
#O[6]:   0.29496

(Estimate the error in the approximation)

#I[7]:: error:(b-a)∧5 D4MAX/(180 n∧4)   (standard error formula for Simpson's
                                          rule, n subdivisions, on [a,b])

#O[7]: $\dfrac{\text{D4MAX} (-a + b)^5}{180\, n^4}$

(Maximize the fourth derivative of g(x) on [0,1])

#I[8]:: D[g[x],{x,4}]

#O[8]:   25Exp[- x^2] Sin[x] + 56x Cos[x] Exp[- x^2] - 72 x^2 Exp[- x^2] Sin[x]

$$- 32 \, x^3 \, \text{Cos}[x] \, \text{Exp}[- \, x^2] + 16 \, x^4 \, \text{Exp}[- \, x^2] \, \text{Sin}[x]$$

#I[9]::   Graph[%,x,0,1]   (low resolution mode)

#O[9]:

```
20 ¦                          **********
   |                        ****      *****
   |                      ***             ***
   |                    ***                ***
   |                  **                    *
   |                **                       **
   |               *                          ***
   |             **                            **
10 ¦            **                              **
   |           *                                 **
   |          **                                  **
   |         *                                      **
   |        **                                       **
   |       *                                           **
   |     **                                             **
   |    *                                                **
 0-------------0.2------------0.4------------0.6------------0.8-----------
   |                                                         **
   |                                                          ***
   |                                                            **
```

#I[10]::   D[g[x],{x,5}]

#O[10]:   81Cos[x] Exp[- x^2] - 250x Exp[- x^2] Sin[x] + ...

#I[11]::   Sol[%=0,x]
#O[11]::   Sol[%=0,x]   (equation can't be solved algebraically)

#I[12]::   <'newton   (load a homemade numerical solver)

#I[13]::   Newton[%,x,.4,.0005]   (''seed'' = 0.4, tolerance = .0005)

zeroth estimate is   0.4
estimate number 1   is   0.40636356
#O[14]:   0.406364

#I[16]::   S[D[g[x],{x,4}],x->.406]   (substitute critical number into function)

#O[16]:   22.736Cos[0.406] Exp[- 0.406^2] + etc.

#I[17]::   N[%]
#O[17]:   20.5887   (This is the desired extremum)

#I[18]::   S[error,{D4MAX->21,a->0,b->1,n->4}]   (substitute for parameters)
#O[18]:   7/15360

```
#I[19]::  N[%]
#O[19]:   4.55729* -4  (estimate .29496 accurate to three decimal places)
```

(Estimate integral using Taylor polynomial)

```
#I[20]::  Ps[g[x],x,0,7]

                  3      5        7
               7 x    27 x    1303 x
#O[20]:    x - ---- + ----- - --------
                6      40       5040

#I[21]::  Int[%,{x,0,1}]
#O[21]:   11633/40320  (closed form integration works easily)

#I[23]::  N[%]
#O[23]:   0.288517  (Compare with Simpson's estimate = .2947)
```